The Uniqueness
of our
Space-time

A Catalogue of Cayley Tables, Division Algebras, and
Emergent Geometric Spaces for most Finite Groups
up to Order Fifteen

by

Dennis Morris

Published by: Abane & Right

31/32 Long Row

Port Mulgrave, Saltburn

TS13 5LF

United Kingdom

01947 840707

Revision January 2016

dennis355@btinternet.com

December 2015

Contents

Preliminaries

The numbers of the finite groups to order 100

G is the number of groups of the given order

AG is the number of abelian groups of the given order

There is a cyclic group of every order, C_n, and so if there is only one group of a given order, say order 85, then that group is a cyclic group. The symmetric groups, S_n, which are groups of permutations, are of order $n!$, and so we have S_3 of order six and S_4 of order twenty-four etc.. The alternating groups, A_n, which are groups of even permutations, are of order $\dfrac{n!}{2}$.

Order	G	AG	Order	G	AG	Order	G	AG	Order	G	AG	Order	G	AG
1	1	1	21	2	1	41	1	1	61	1	1	81	15	5
2	1	1	22	2	1	42	5	1	62	2	1	82	2	1
3	1	1	23	1	1	43	1	1	63	4	3	83	1	1
4	2	2	24	15	3	44	4	2	64	267	11	84	15	2
5	1	1	25	2	2	45	2	2	65	1	1	85	1	1
6	2	1	26	2	1	46	2	1	66	4	1	86	2	1
7	1	1	27	5	3	47	1	1	67	1	1	87	1	1
8	5	3	28	4	2	48	52	5	68	5	2	88	12	3
9	2	2	29	1	1	49	2	2	69	1	1	89	1	1
10	2	1	30	4	1	50	5	2	70	4	1	90	10	2
11	1	1	31	1	1	51	1	1	71	1	1	91	1	1
12	5	2	32	51	7	52	5	2	72	50	6	92	4	2
13	1	1	33	1	1	53	1	1	73	1	1	93	2	1
14	2	1	34	2	1	54	15	3	74	2	1	94	2	1
15	1	1	35	1	1	55	2	1	75	3	2	95	1	1
16	14	5	36	14	4	56	13	3	76	4	2	96	230	7
17	1	1	37	1	1	57	2	1	77	1	1	97	1	1
18	5	2	38	2	1	58	2	1	78	6	1	98	5	2
19	1	1	39	2	1	59	1	1	79	1	1	99	2	2
20	2	2	40	14	3	60	13	2	80	52	5	100	16	4

The simple finite groups (not all groups are simple) are either: 1) Cyclic groups of prime order 2) Alternating groups of degree five or more 3) Chevalley groups 4) Twisted Chevalley groups 5) one of the 28 sporadic groups.

Some groups have more than one name. In these cases, the most commonly used appellation is presented before the isomorphism sign.

The C stands for cyclic and C_n is the cyclic group of order n.

The S stands for symmetric. For these groups, the subscript is not the order of the group.

The D stands for dihedral. For these groups, the subscript is not the order of the group.

Q stands for the so called quaternion group sometimes written as the dicyclic group Q_4 .

The A stands for Alternating. For these groups, the subscript is not the order of the group.

T is the cubic group which is also known as the dicyclic group Q_6 .

Order	Abelian groups	Non-abelian groups
1	C_1	
2	$C_2 \approx S_2$	
3	C_3	
4	C_4, $C_2 \times C_2$	
5	C_5	
6	C_6	$S_3 \cong D_3$
7	C_7	
8	C_8, $C_2 \times C_4$, $C_2 \times C_2 \times C_2$	D_4, Q
9	C_9, $C_3 \times C_3$	
10	C_{10}	D_5
11	C_{11}	
12	C_{12}, $C_2 \times C_6$	A_4, D_6, $Q_6 \cong T$
13	C_{13}	
14	C_{14}	D_7
15	C_{15}	

There are fourteen order 16 groups.

Proper Subgroups to Order 15

If a group appears alone, like C_5, then there is only one subgroup of that form within the given group. Where there are multiple subgroups of the same form, they are shown like $3(C_2)$.

Order	Group	Proper subgroups
1	C_1	-
2	$C_2 \cong S_2$	-
3	C_3	-
4	C_4	C_2
4	$C_2 \times C_2$	$3(C_2)$
5	C_5	-
6	C_6	C_2, C_3
6	$S_3 \cong D_3$	$3(C_2)$, C_3
7	C_7	-
8	C_8	C_2, C_4
8	$C_2 \times C_4$	$3(C_2)$, $2(C_4)$, $C_2 \times C_2$
8	$C_2 \times C_2 \times C_2$	$7(C_2)$, $7(C_2 \times C_2)$
8	D_4	C_4, $5(C_2)$, $2(C_2 \times C_2)$
8	Q	C_2, $3(C_4)$
9	C_9	C_3
9	$C_3 \times C_3$	$4(C_3)$
10	C_{10}	C_2, C_5
10	D_5	$5(C_2)$, C_5
11	C_{11}	-
12	C_{12}	C_2, C_3, C_4, C_6
12	$C_2 \times C_6$	$3(C_2)$, C_3, $C_2 \times C_2$, $3(C_6)$
12	A_4	$3(C_2)$, $C_2 \times C_2$, $4(C_3)$
12	D_6	$7(C_2)$, C_3, C_6, $3(C_2 \times C_2)$, $2(D_3)$
12	$T \cong Q_6$	C_2, C_3, $3(C_4)$, C_6
13	C_{13}	-
14	C_{14}	C_2, C_7
14	D_7	$7(C_2)$, C_7
15	C_{15}	C_3, C_5

SECTION I – Introduction

Preliminary Note

The result of the work presented in this book is that:

There are no geometric spaces within the finite groups

other than the two 4-dimensional geometric spaces which together

form our observed universe.

We feel it cannot be just coincidence that the only geometric spaces within the finite groups exactly match our observed universe, for it is infinitely unlikely that, of all the many infinities of finite groups, the only one that holds geometric spaces holds exactly the geometric spaces which we observe in our universe and no more geometric spaces than those geometric spaces which we observe in our universe.

Other than the finite group $C_2 \times C_2$, any group which holds a geometric space must be of the following nature:

a) Must be of order 2^n : $n > 2$
b) Must not be C_4.
c) Must have no C_4 subgroup.

These three conditions, only two really, are (Sylow's theorems) sufficient to prove only the $C_2 \times C_2$ finite group holds a geometric space. We might add weight by mentioning other conditions.

d) Must have no proper cyclic subgroups of order fifteen or less other than the order two cyclic group C_2
e) Must have no proper subgroups of prime order other than the order two cyclic group C_2.
f) Must have no $C_2 \times C_2 \times C_2$ subgroup; must have no $C_2 \times C_4$ subgroup; must have no Q subgroup; must have no D_4 subgroup: must have no S_3 subgroup.

The dimension of the geometric space within a finite group must be the same as the order of the finite group. Since we observe about us only 4-dimensional space, then, by observation, we might assert that no finite group holds geometric spaces except the order four groups. We know that, of the order four groups, only the $C_2 \times C_2$ group holds geometric spaces. We have an observational proof of the statement of the first sentence.

This book presents what we have and explains what we mean by a geometric space and how the two geometric spaces we observe in our universe are held within the finite group $C_2 \times C_2$.

The Higher Dimensional Division Algebras in General

A division algebra is a type of numbers. Examples of division algebras are the real numbers, $\mathbb{R}$, the Euclidean complex numbers, $\mathbb{C}$, the hyperbolic complex numbers (2-dimensional space-time), $\mathbb{S}$, and the quaternions. There are very many different types of division algebras; there might well be an infinite number of different types of division algebras.

Note: When we use the term 'division algebra', we drop the requirement for additive inverses on the real axis[1]. Otherwise, we comply with every standard division algebra axiom.

The division algebras are within the finite groups. It seems that each finite group holds at least one type of division algebra. The order of the finite group is the dimension of the division algebra; for example; the 1-dimensional real numbers are within the order one group C_1; the 2-dimensional Euclidean complex numbers are within the order two group C_2; the 2-dimensional hyperbolic complex numbers are also within the group C_2; the 4-dimensional quaternions are within the order four group $C_2 \times C_2$. The separate algebras within a given group are expressed as the algebraic matrix form of that group; for example, the algebraic matrix form of the C_2 group is:

$$\exp\left(\begin{bmatrix} a & b \\ P_{2,1}b & a \end{bmatrix}\right) \tag{1.1}$$

By setting the parameter, $P_{2,1}$, to plus unity and to minus unity, we get the two C_2 division algebras.

$$\mathbb{S} = \begin{bmatrix} e^a & 0 \\ 0 & e^a \end{bmatrix}\begin{bmatrix} \cosh b & \sinh b \\ \sinh b & \cosh b \end{bmatrix} \quad \& \quad \mathbb{C} = \begin{bmatrix} e^a & 0 \\ 0 & e^a \end{bmatrix}\begin{bmatrix} \cos b & \sin b \\ -\sin b & \cos b \end{bmatrix} \tag{1.2}$$

In fact, the parameter, $P_{2,1}$, can be any non-zero value.

Although some of the division algebras such as the Euclidean complex numbers, the hyperbolic complex numbers, or the quaternions have been known for a hundred years or more, and the real numbers have been known for many thousands of years, the other division algebras have been discovered only recently. There is much not known about these recently discovered division algebras and about the number of them and the distribution of them. We list a few of the unknowns in this area of mathematics:

    a) Does every finite group hold division algebras within it?

---

[1] This dropping of additive inverses on the real axis corresponds to being unable to travel backwards in time in the 2-dimensional case. Since this fits with observation, and with the maths, we opine that insistence upon additive inverses on the real axis is piety rather than science. We do have additive inverses in every other axis.

Since the first discovery of these recently discovered division algebras (published 2007[2]), it was tacitly assumed that every finite group holds many non-isomorphic division algebras. It is certainly true that every finite group holds at least one division algebra, and we now know that the lowest order dicyclic group, also called the quaternion group, holds only one non-isomorphic type of division algebra. However, it is sometimes very difficult to find the algebraic matrix form of a group, and we need the algebraic matrix form to have the different division algebras if there is more than one type of non-isomorphic algebra. Although, in some cases, the algebraic matrix form is eventually found, it is not understood why previous only slightly different attempts failed. Such failures are usually attributed to human error, but we have no proof that the algebraic matrix form exists until we find it. There are low order groups whose algebraic matrix form is unknown to date; the groups $D_5 \& Q_6$ are examples[3]. There is a lower order finite group, A_4 for which we have an algebraic matrix form, but we are not confident that we have done the calculation correctly.

b) How many separate (many will be isomorphic) division algebras does a given finite group hold?

Up to order fifteen, the cyclic groups follow a pattern in which the numbers of separate algebras is simply $2^{(Order-1)}$. Because we do not properly understand the formulation of the algebraic matrix form for even the cyclic groups, we cannot be sure this pattern will continue. Most other low order groups have $2^{(Order-1)}$ separate division algebras, but there are groups like $C_2 \times C_2$, which have 2^{Order} or more division algebras within them. The number of separate algebras is not simply $2^{Number\ of\ Parameters}$, and so we will not necessarily know the number of separate algebras if we know the number of parameters in the algebraic matrix form. Since the formation of the algebraic matrix form is not properly understood, we do not know with certainty how many parameters a given group will have in its algebraic matrix form until we find that algebraic matrix form.

c) How many non-isomorphic division algebras will a group hold?

Again, up to order fifteen, the cyclic groups follow a pattern based upon their subgroup structure (the numbers of group elements of a given order). In particular, the prime order cyclic groups always have (up to order fifteen) the same number of non-isomorphic division algebras as the order of the group. The number of a particular set of isomorphic division algebras within a group again seems predictable within a given cyclic group, but the prediction is poorly founded since we do not clearly understand the way the algebraic matrix form is produced.

d) How many of the finite groups hold a geometric space?

Within this book, your author records his search for geometric spaces within the finite groups. That search, covering only a small number of low order finite groups, finds two geometric spaces within the group $C_2 \times C_2$ but no other group that holds a geometric space. The search enables us to show that the vast majority of finite groups of higher orders do not hold a geometric space, but not all groups of higher orders are necessarily caught by the results of the search. Because the two geometric spaces within the group $C_2 \times C_2$ together are the 4-dimensional space-time and the quaternion geometric space of quantum physics which we observe in the universe around us, we opine that there can be no more geometric spaces

---

[2] Dennis Morris: Complex Numbers The Higher Dimensional Forms. There is now a second edition available.

[3] The algebraic matrix form of the order twelve dicyclic group Q_6 has now been found.

within the finite groups for, if there were such other geometric spaces, we would observe them. However, this is poorly understood, and an opinion is not a mathematical proof.

At the centre of our lack of understanding is that we do not know at more than a 'guess and try it' calculation level how the algebraic matrix form of a group is derived. Your author's opinion is that a specialist group theorist might be able to untangle this problem, but your author admits he lacks the specialist knowledge to achieve this goal.

The first step in untangling any mathematical problem is to gather data. We might begin by noticing a pattern in something; for example, we might notice that twice the number 142857 has the same digits in the same order;

$$2 \times 142857 = 285714 \tag{1.3}$$

We immediately wonder if three times this number is also special, and so we test it; we gather data:

$$3 \times 142857 = 428571 \tag{1.4}$$

We continue to gather data and find the pattern repeating until something unexpected happens:

$$7 \times 142857 = 999999 \tag{1.5}$$

We now have enough data to realise something is happening here. Perhaps we see into the maths at this point, or perhaps we seek more data. We discover:

$$2 \times 0588235294117647 = 1176470588235294$$
$$17 \times 0588235294117647 = 9999999999999999 \tag{1.6}$$

Perhaps we then notice that both 7 and 17 are prime numbers and that:

$$\frac{1}{7} = 0.(142857)_{\text{Recurring}} \tag{1.7}$$

We have our insight. The secret is out.

This book is a first step into exploring the division algebras within the finite groups. We gather data, but we do not yet have the insight we need. Your author's interests are in physics, and so he will not pursue this area of maths any deeper at present. This book is written to provide other interested mathematicians with the first bits of data. Perhaps this book will be a first step for a group theorist.

Chapter 2

An Overview of Geometric Space

This book assumes some familiarity with the spaces that derive from the finite groups as division algebras. If the reader is unfamiliar with these division algebras, then reading the chapter titled 'The Lower Order Groups' will enlighten the reader.

Examples of such division algebra spaces are the 2-dimensional Euclidean complex plane of the 2-dimensional Euclidean complex numbers, $\mathbb{C}$, the 2-dimensional space-time of the hyperbolic complex numbers, $\mathbb{S}$, or the 4-dimensional quaternion space. Such spaces are called spinor spaces or division algebra spaces. If the reader is unfamiliar with these division algebras and the associated complex spaces (spinor spaces), then knowledge of these can be gained from the books by your author listed at the end of this book.

These division algebra spaces are not geometric spaces. A geometric space is a thing very different from a division algebra space. The geometric spaces emerge from the division algebra spaces by superimposition of a complete set of algebraically isomorphic division algebra spaces. An example of a geometric space is our 4-dimensional space-time.

This book, as far as it goes, documents the search through the lower order finite groups for all the different types of geometric space, such as our 4-dimensional space-time, which emerge, via the division algebra spaces, from the finite groups. The structures, geometric or otherwise, of these emergent spaces are determined by the nature of the emergent distance functions, and it is for these emergent distance functions that we seek. The search is done by brute force calculation using computers. There are an infinite number of groups, and it is not known how to discover such emergent distance functions analytically. Since the search does not cover all the finite groups, the search is incomplete. The result of the search is that there are only two types of geometric space of low dimension. Your author has previously derived the nature of the 4-dimensional space-time of our universe[4] from the finite groups.

The search documented in this book gives weight to the unproven proposition that our universe is unique, and that there are no geometric spaces of dimension other than four.

The nature of geometric space:
Our 4-dimensional space-time is not a division algebra space; it is not a spinor space. The spinor spaces are division algebras and as such, the norm (distance function) of these spaces is such that its form is maintained under multiplication. For example we see the complex numbers, $\mathbb{C}$, have distance function given by the determinant of the algebraic matrix form of the division algebra:

---

[4] See: Upon General Relativity by Dennis Morris

$$a+ib \equiv \begin{bmatrix} a & b \\ -b & a \end{bmatrix}, \qquad \exp\left(\begin{bmatrix} a & b \\ -b & a \end{bmatrix}\right) = \begin{bmatrix} r & 0 \\ 0 & r \end{bmatrix}\begin{bmatrix} \cos b & \sin b \\ -\sin b & \cos b \end{bmatrix} \quad : \quad r = e^{a}$$

$$\det\left(\begin{bmatrix} a & b \\ -b & a \end{bmatrix}\right) \equiv \det\left(\begin{bmatrix} r & 0 \\ 0 & r \end{bmatrix}\begin{bmatrix} \cos b & \sin b \\ -\sin b & \cos b \end{bmatrix}\right) \tag{2.1}$$

$$a^2 + b^2 \equiv r^2$$

Since the form of the algebraic matrix form is closed under multiplication (this is multiplicative closure) the form of the determinant is closed under multiplication. We have:

$$\left(a^2+b^2\right)\left(c^2+d^2\right) = \left(ac-bd\right)^2 + \left(ad+bc\right)^2 \tag{2.2}$$

The distance function of our 4-dimensional space-time does not have this property of multiplicatively closed form:

$$\left(t^2-x^2-y^2-z^2\right)\left(a^2-b^2-c^2-d^2\right) \neq P^2 - Q^2 - R^2 - S^2 \tag{2.3}$$

Our 4-dimensional space-time is a geometric space, also called a fabricated space, rather than a division algebra space. Note that 2-dimensional space-time is a spinor space.

For the elucidation of the reader, we will compare our 4-dimensional space-time with the 4-dimensional quaternion space. Our 4-dimensional space-time is a geometric space while quaternion space is a spinor space, and so we are comparing the nature of a geometric space with the nature of a spinor space. Each of these two spaces has four variables corresponding to the four dimensions. The distance function (norm) of quaternion space is:

$$d^4 = \left(a^2+b^2+c^2+d^2\right)^2$$
$$d^2 = a^2+b^2+c^2+d^2 \tag{2.4}$$

Within our 4-dimensional space-time, all four variables are real variables, $\in \mathbb{R}$. Within our 4-dimensional space-time, we have 2-dimensional rotation in six planes, three Euclidean planes and three space-time planes also known as Lorentz boosts. Within our 4-dimensional space-time, all possible pairings of two variables form a plane in which we can rotate.

Within quaternion space, because the quaternions are a division algebra, there is one real variable and three imaginary variables. Within quaternion space, since there are three 2-dimensional sub-algebras, there is rotation in only the three 2-dimensional planes formed by the real variable and each of the imaginary variables. In quaternion space, any two imaginary variables are not connected by 2-dimensional rotation because two imaginary variables alone do not form a division algebra. The quaternion space is constrained by its algebraic properties.

Because our 4-dimensional space-time, having real variables, is not a division algebra space, it is not constrained by algebraic properties.

Looking at the distance function of quaternion space, (2.4), we see that putting any two variables to zero would give the distance function of a 2-dimensional Euclidean space. If the variables in quaternion space were real, then there would be no constraints on the variables and we would have a 4-dimensional geometric space with six 2-dimensional Euclidean rotation planes. Such a space does exist; it emerges from the $C_2 \times C_2$ finite group by superimposition of the two quaternion algebras (the quaternions and the anti-quaternions). Actual quaternion space is a single spinor space; it is not the superimposition of two spinor spaces.

In general, the division algebra spaces of dimension greater than two, being division algebras, are such that there is no rotation between all pairs (or triples or…) of imaginary variables.

Now try to visualise a 3-dimensional quaternion type of space formed from one real variable and two imaginary variables and in which there are two 2-dimensional rotation planes each formed from the real variable and one of the imaginary variables. No such space exists, but it is easier to visualise than the 4-dimensional quaternion space. Now ask yourself this question. Without rotation in the third 2-dimensional plane of the two imaginary variables, can such a space have a geometrical spatial structure? Would you be able to wave your arms around in it? Imagine that the horizontal plane is the absent rotation plane in such a space. By rotating in both vertical planes at right-angles to each other, your hand could visit only four points on the horizontal 'circle' with centre your elbow, each point being at 90^0 to its neighbours. (Remember, you cannot rotate the co-ordinate grid in the horizontal plane.) The horizontal plane, without rotation, would be quantitised. Your author opines that such a space cannot be a space in which he can wave his arms freely. By this he means that such a space is not a geometric space. A geometric space is a space in which your author can wave his arms around freely. A geometric space must have real variables.

Your author asserts that, because a geometric space must have real variables, it can arise from only the superimposition of isomorphic division algebras.[5] We know how to find all the division algebras; they are within the finite groups. The reader might assume that we can superimpose the isomorphic division algebras and thereby derive every possible geometric space. The reader is correct, but, seemingly, there are only two geometric spaces.

Many of the aspirant geometric spaces which emerge from the superimposition of isomorphic algebras have emergent distance functions which can support rotation between only some pairs of variables. By far the greatest majority of aspirant geometric spaces simply cannot hold 2-dimensional rotation at all because the emergent distance function will not factorise into quadratic form.

Further, being able to support rotation between all pairs of variables, although a necessary condition, is not a sufficient condition to form a geometric space. Some of the aspirant geometric spaces which emerge from the superimposition of isomorphic algebras have emergent distance functions which can support rotation between all possible pairs of variables but cannot form a geometric space of more than 2-dimensions. An example is a space in which the all pairings of variables are of a space-time nature. Consider the three pairings:

$$dist^2 = a^2 - b^2 \qquad dist^2 = c^2 - d^2 \qquad dist^2 = e^2 - f^2 \qquad (2.5)$$

Because there are three pairs of variables but only two possible signs (plus and minus), we cannot form a 3-dimensional distance function with space-time rotation in all three planes.

---

[5] See: Dennis Morris *Upon General Relativity*.

For sometimes different reasons, most finite groups, and seemingly all but one finite group, do not give rise to a geometric space. Seemingly, the two geometric spaces which do emerge from that one group together form our universe.

In this book, we are concerned with only 2-dimensional rotations. We have searched for geometric spaces which hold higher dimensional rotations, and, other than the quaternion emergent space, we have found none within the finite groups considered in this book.

Clearly, the reader needs to understand how we calculate an emergent distance function. We begin this book with the calculation of the emergent distance functions of the only two known emergent geometric spaces. Your author opines that this is the simplest way to introduce the reader to the concept of an emergent distance function.

The bulk of this book:

Although the prime motivation for this book is to seek geometric spaces within the finite groups, since there is, at this time, no known means of discovering such geometric spaces other than by the brute force of thousands of individual calculations, along the way, since it has never been done before, we might as well catalogue all the division algebras with which we have to deal. Similarly, although it has been done before, we might as well catalogue the Cayley tables of the finite groups with which we deal. We use the Standard form Cayley tables of each group. Such presentation is both enlightening and useful. We anticipate that such a catalogue will be of some time-saving practical use to many mathematicians.

Calculating an Emergent Distance Function

There are two ways in which we might calculate an emergent distance function for a given set of isomorphic division algebras. We call the first calculation method *Reduce then Sum* and we call the second calculation method *Sum then Reduce*. The two calculations do not give the same emergent distance function.

We begin by stating that, although, we do not understand with clarity which of the two calculation methods we should use to calculate the emergent distance function, experience of and the results of the computer calculations presented in this book are sufficient to convince us that the *Reduce the Sum* method is the correct method to use. We can state with certainty that there are no *Sum then Reduce* emergent distance functions which support a geometric space within the finite groups considered in this book. We are happy to dismiss the *Sum then Reduce* calculation as incorrect, but we owe the reader a look at this calculation method, and so we present the reader with both methods for completeness.

The calculation of an emergent distance function:
Each type of division algebra has associated with it an emergent distance function which determines the nature of the emergent space.

The commutative $C_2 \times C_2$ finite group contains sixteen division algebras. Of these sixteen division algebras, eight are non-commutative division algebras. Of these eight non-commutative division algebras, six are the A_3 algebras. The six A_3 algebras are algebraically isomorphic to each other; they differ from each other only by the basis in which they are written. None-the-less, the finite group $C_2 \times C_2$ contains all six A_3 algebras and it seems that reality uses all six A_3 algebras.

The distance functions of the six A_3 algebras are the determinants of the six A_3 algebraic matrix forms:

$$d^4{}_{SSA1} = t^4 + x^4 + y^4 + z^4 + 2\left(-t^2x^2 - t^2y^2 + t^2z^2 + x^2y^2 - x^2z^2 - y^2z^2\right)$$

$$d^4{}_{SSA2} = t^4 + x^4 + y^4 + z^4 + 2\left(-t^2x^2 - t^2y^2 + t^2z^2 + x^2y^2 - x^2z^2 - y^2z^2\right)$$

$$d^4{}_{SAS1} = t^4 + x^4 + y^4 + z^4 + 2\left(-t^2x^2 + t^2y^2 - t^2z^2 - x^2y^2 + x^2z^2 - y^2z^2\right)$$

$$d^4{}_{SAS2} = t^4 + x^4 + y^4 + z^4 + 2\left(-t^2x^2 + t^2y^2 - t^2z^2 - x^2y^2 + x^2z^2 - y^2z^2\right)$$

$$d^4{}_{ASS1} = t^4 + x^4 + y^4 + z^4 + 2\left(t^2x^2 - t^2y^2 - t^2z^2 - x^2y^2 - x^2z^2 + y^2z^2\right)$$

$$d^4{}_{ASS1} = t^4 + x^4 + y^4 + z^4 + 2\left(t^2x^2 - t^2y^2 - t^2z^2 - x^2y^2 - x^2z^2 + y^2z^2\right)$$

(3.1)

The A_3 emergent space will be 4-dimensional because there are four variables in the emergent distance function.

The A_3 emergent space will support 2-dimensional Euclidean rotations between two of its variables if the emergent distance function is of the form $d^2 = x^2 + y^2$ when the other two of the variables are zero, and it will support 2-dimensional space-time rotations between two of its variables if the emergent distance function is of the form $d^2 = x^2 - y^2$ when the other two of the variables are zero. The conditions are a little less constraining than this because we can accept a numerical factor in the space-time case. Since time and space are measured in different units, we need only that the emergent distance function is of the form $d^2 = Nx^2 - y^2$, $N \in \mathbb{R}$, when the other two of the variables are zero. The N can be absorbed in the units.

An incorrect way to proceed:
We opine that this is the incorrect way to proceed. We call this the *Sum then Reduce* method of forming the emergent distance function. We form the *Sum then Reduce* A_3 emergent distance function by first summing the six individual A_3 distance functions, (3.1) and then reducing this sum by factorisation to quadratic form. Adding the individual A_3 distance functions, (3.1) gives:

$$6d^4_{\;EEDF} = 6t^4 + 6x^4 + 6y^4 + 6z^4 - 4\left(t^2x^2 + t^2y^2 + t^2z^2 + x^2y^2 + x^2z^2 + y^2z^2\right) \qquad (3.2)$$

The computer tells us that this will not factorise (reduce) to a quadratic form, and so the *Sum then Reduce* method will not give us a emergent distance function which supports 2-dimensional rotations. Hang on! the reader protests, just because the computer can't factorise (3.2) does not mean it will not factorise. Well, if (3.2) factorises when all variables are non-zero, it will certainly factorise when all but two of the variables are zero.

If we can show that an expression of many variables will not factorise when all but two of the variables are zero, then we have proven that the expression will not factorise.

For our purposes, we need the emergent distance function to factorise into a quadratic form that will support one or both types of 2-dimensional rotation. (2-dimensional rotation respects distance functions of the quadratic form $dist^2 = ...$)

Setting two variables, $\{y, z\}$, of (3.2) to zero gives:

$$d^4_{\;EEDF} = t^4 + x^4 - \frac{2}{3}t^2x^2 \qquad (3.3)$$

Will this reduce to the form of one of the two possible 2-dimensional distance functions? We see immediately that this cannot be of the form of a Euclidean 2-dimensional distance function:

$$d^2 = t^2 + x^2$$
$$d^4 = \left(t^2 + x^2\right)^2 = t^4 + x^4 + 2x^2t^2 \qquad (3.4)$$

The 2-dimensional space-time form is:

$$d^2 = A^2 t^2 - x^2$$
$$d^4 = \left(A^2 t^2 - x^2\right)^2 = A^4 t^4 + x^4 - 2A^2 x^2 t^2 \tag{3.5}$$

We need a value for A such that:

$$A^4 t^4 + x^4 - 2A^2 x^2 t^2 = t^4 + x^4 - \frac{2}{3} x^2 t^2 \tag{3.6}$$

Looking at the t^4 term, we see that we need $A^2 = 1$ and looking at the $x^2 t^2$ term, we see that we need $A^2 = \frac{1}{3}$. Clearly, we do not have a 2-dimensional space-time rotation supported by the emergent distance function (3.2).

The mathematics given above, (3.2) to (3.6), which we call the *Sum then Reduce* method is, we opine, the wrong way to approach the calculation of an emergent distance function. This method has been tested over the finite groups in this book and does not produce any geometric spaces.

The correct way to proceed:
We think this is the correct way to proceed. We call this the *Reduce then Sum* method of forming the emergent distance function. Suppose we first reduce the distance functions, (3.1), to their quadratic equivalent form. This means we must take the square roots of both sides of the A_3 distance functions, (3.1). In order to support 2-dimensional rotations, the above distance functions, (3.1), must factorise into the square of an expression which supports 2-dimensional rotations.

In general, the power of the individual distance functions of a division algebra is equal to the order of the underlying group because the individual distance function of a division algebra is the determinant of a square matrix of size equal to the order of the group. Such distance functions cannot be reduced to a quadratic form unless the order of the underlying group is divisible by two. Similarly, such distance functions cannot be reduced to a cubic form which might support 3-dimensional rotation unless the order of the underlying group is divisible by three, and so on for higher dimensional types of rotation. It does not follow that a division algebra from a group of even order can be factorised to a quadratic form; the condition is necessary but not sufficient.

Of course, and this is particularly relevant to high dimensional algebras, our inability to factorise an expression into the square of some expression does not prove that the expression cannot be so factorised. We have illustrated above, (3.3) to (3.6), how we might prove non-factorisation by considering only two non-zero variables. In the A_3 case, the distance functions reduce to:

$$d^4{}_{SSA1} = \left(t^2 - x^2 - y^2 + z^2\right)^2$$

$$d^4{}_{SSA2} = \left(t^2 - x^2 - y^2 + z^2\right)^2$$

$$d^4{}_{SAS1} = \left(t^2 - x^2 + y^2 - z^2\right)^2$$

$$d^4{}_{SAS2} = \left(t^2 - x^2 + y^2 - z^2\right)^2 \qquad (3.7)$$

$$d^4{}_{ASS1} = \left(t^2 + x^2 - y^2 - z^2\right)^2$$

$$d^4{}_{ASS2} = \left(t^2 + x^2 - y^2 - z^2\right)^2$$

Taking the square roots of both sides gives the 2-dimensional equivalent quadratic distance functions. We have now done the 'reduce' part of the *Reduce then Sum* method.

We form the 2-dimensional emergent distance function by summing the 2-dimensional equivalent quadratic distance functions. This is:

$$6d^2{}_{Emerg} = sum \begin{Bmatrix} t^2 - x^2 - y^2 + z^2 \\ t^2 - x^2 - y^2 + z^2 \\ t^2 - x^2 + y^2 - z^2 \\ t^2 - x^2 + y^2 - z^2 \\ t^2 + x^2 - y^2 - z^2 \\ t^2 + x^2 - y^2 - z^2 \end{Bmatrix} = 6t^2 - 2x^2 - 2y^2 - 2z^2 \qquad (3.8)$$

This gives:

$$3d^2{}_{Emerg} = 3t^2 - x^2 - y^2 - z^2 \qquad (3.9)$$

We ignore the 3 on the LHS because it is just a scaling factor. We can ignore the 3 coefficient of the t^2 because we can adjust the units in which we measure time. This gives the 2-dimensional A_3 emergent distance function as:

$$d^2 = t^2 - x^2 - y^2 - z^2 \qquad (3.10)$$

Setting the six different permutations of two variables to zero gives us six 2-dimensional distance functions (three of each type) respected by the two different types of 2-dimensional rotation.

$$\begin{array}{ccc} d^2 = t^2 - x^2 & d^2 = t^2 - y^2 & d^2 = t^2 - z^2 \\ d^2 = x^2 + y^2 & d^2 = x^2 + z^2 & d^2 = y^2 + z^2 \end{array} \qquad (3.11)$$

We see that every variable 'hangs together' with every other variable in a 2-dimensional distance function; by this, we mean that there is 2-dimensional rotation connecting every pair of variables and thereby does the emergent space have a geometric structure.

This, (3.10), is our 4-dimensional space-time.

The other geometric space:

It is conjectured that there is only one other emergent distance function with the requisite properties necessary to form a geometric space. This geometric space is from the other two non-commutative $C_2 \times C_2$ algebras which are the quaternions and the anti-quaternions. That 2-dimensional emergent distance function is:

$$2d^2_{\:Emerg} = sum \begin{Bmatrix} t^2 + x^2 + y^2 + z^2 \\ t^2 + x^2 + y^2 + z^2 \end{Bmatrix} \tag{3.12}$$

$$d^2 = t^2 + x^2 + y^2 + z^2$$

Why bother?:

It seems very likely that the whole of theoretical physics other than the strong force is contained within the eight non-commutative A_3 algebras. Perhaps these emergent distance functions are unique or perhaps there are others connected to the strong force.

What about 3-dimensional rotations?:

Is there a 6-dimensional emergent distance function which 'hangs together' and supports 3-dimensional rotations? If the reader looks ahead to the order three group algebras, the reader will see that the 3-dimensional division algebras all have distance functions of the form, up to signs:

$$d^3 = a^3 + b^3 + c^3 - 3abc \tag{3.13}$$

A 6-dimensional distance function which supported 3-dimensional rotations between all twenty triples of variables would have to be of the form:

$$dist^{6.} = \left(a^3 + b^3 + c^3 + d^3 + e^3 + f^3 - 3abc - 3abd - 3acd - 3bcd - ...3cdf - ... - 3def \right)^2 \tag{3.14}$$

There are twenty $-3adf$ type terms. We have looked at the order six groups to see if such an emergent distance function arises – it does not.

Chapter 4

More Introduction and Some Details

Based upon each finite group are a number of division algebras of dimension equal to the order of the group. These division algebras are presented and catalogued as their algebraic matrix forms from which the individual algebra can be calculated by selecting values for the given parameters and taking the exponential of the resultant matrix. Of these division algebras, many are algebraically isomorphic. The process of calculating an individual algebra is presented to the reader using the low order groups.

The division algebras each have a distance function (norm). These distance functions are catalogued only for the low order groups (low dimensional algebras) because these norms become very long and unmanageable for higher order groups. Such norms can be calculated by computer using the given algebraic matrix form of the algebra. Such norms are most easily handled by computer.

Notes:
We will denote the identity of the group with the Latin letter a.

The Cayley tables are presented in standard form; this is the format with the identity on the leading diagonal. The standard form, seen as a matrix with each group element seen as a real variable, is the basis of the algebraic matrix form of the group. Other forms of the Cayley tables can be found by swapping rows or swapping columns of the Standard form Cayley table.

The top row of the Standard form Cayley table and the leftmost column of the Standard form Cayley table are direct copies of the labels of the group elements, and so these labels are not presented above or beside the Cayley table.

The standard form of the Cayley table is such that it can be separated into its different parts (different variables) to form the set of group elements. If, after separating, the variables are all set to unity, we are left with the set of permutation matrices which correspond one-to-one with the elements of the group.

The Standard form Cayley table is useful for calculating the subgroup structure of the group. Taking any one of the individual permutation matrices and continually multiplying it by itself until it equals the identity on the leading diagonal will give the order of that element and thereby produce a list of all elements in a subgroup. In technical terms, we are using the permutation matrix as a generator of the subgroup.

There is, except for the order one group always more than one division algebra within each finite group although these separate division algebras might be isomorphic. The individual division algebras are found by inserting parameters into the algebraic matrix form of the group and subsequently eliminating some of these parameters by insisting upon multiplicative closure of the algebraic matrix form under matrix multiplication. Multiplicative closure usually, but not always, arises when all but $(n-1)$ parameters have been eliminated where n is the order of the finite group. Having obtained the multiplicatively closed algebraic matrix form, each possible permutation of the parameters each set to ± 1 gives a division algebra matrix form; there are 2^{n-1} such permutations.

The actual division algebra is then formed by taking the exponential of the division algebra matrix form. We give examples of this in the text.

Clarification of the above:

Other than the next chapter, the first part of this book deals with the finite groups up to and including order 4. Within this first part, there is much explanation and examples of the above, and the reader is urged to read this first part of the book to gain a deeper understanding of the above.

Chapter 5

Emergent Spaces and 2-Dimensional Algebras

There are only two possible 2-dimensional spinor rotations. A spinor rotation is rotation within a division algebra[6]. The two possible 2-dimensional spinor rotations are the Euclidean rotation of the 2-dimensional plane of the complex numbers, $\mathbb{C}$, which respects the distance function $d^2 = x^2 + y^2$ and the space-time rotation of the plane of the hyperbolic complex numbers, $\mathbb{S}$, which respects the distance function $d^2 = t^2 - z^2$.

If two variables of an emergent distance function do not 'hang together' in either of these ways, then that emergent distance function does not form a geometric space with 2-dimensional rotations. This does not mean the emergent space is not a geometric space; it might be a geometric space of 3-dimensional rotations[7] or a geometric space of 4-dimensional rotations… However, for the time being, we will be concerned with only 2-dimensional rotations.

A higher dimensional emergent space with 2-dimensional rotations must have an emergent distance function that accommodates one or both of the 2-dimensional distance functions between all pairs of variables. Thus, the distance function of the higher dimensional emergent space must reduce to a quadratic form with as many variables as there are dimensions in the emergent space. An example of such a quadratic form is:

$$d^2 = t^2 + v^2 + w^2 + x^2 - y^2 - z^2 \qquad (5.1)$$

With thought, the reader will realise that, if we are to have 2-dimensional rotations connecting each pair of variables in a space, then quadratic forms of the same nature as (5.1) are the only possibilities. Thus it is that we will be easily able to recognise the emergent distance function of a geometric space that supports 2-dimensional rotations. Essentially, what we are doing in this book is looking within the finite groups for emergent distance functions that are quadratic forms like (5.1). Perhaps we should emphasize the previous sentence.

This book documents the search within the finite groups for emergent distance functions that are quadratic forms like (5.1).

The quadratic form will have a number of plus signs and a number of minus signs; these numbers will dictate how many 2-dimensional planes (pairs of variables) have Euclidean rotation and how many 2-dimensional planes (pairs of variables) have space-time rotation (Lorentz boost).

In 2-dimensions, we have four possible ways to put two variables into a quadratic form distance function. The two squares of the variables can either have the same sign, which is the conserved distance function of a Euclidean rotation, or the two squares of the variables can have different signs, which is the conserved distance function of a 2-dimensional space-time rotation. These four possibilities are:

---

[6] Spinor rotations are n-dimensional rotations in n-dimensional spaces; they are not rotations about an axis.
[7] We will meet 3-dimensional rotations later.

$$d^2 = x^2 + y^2$$
$$d^2 = t^2 - z^2$$
$$d^2 = -x^2 - y^2$$
$$d^2 = -t^2 + z^2$$

(5.2)

In fact, there are only two possible forms here. One form has all the signs the same (either all pluses or all minuses); the other form has two different signs. Thus, we need to consider only:

$$d^2 = x^2 + y^2$$
$$d^2 = t^2 - z^2$$

(5.3)

In n-dimensional space, there are n variables in the distance function. These n variables can be paired together to form 2-dimensional planes in $(n-1)+(n-2)+...+2+1 = \dfrac{n(n-1)}{2}$ ways[8].

In 3-dimensions, the quadratic form distance function is of 2 possible forms:

$$d^2 = t^2 + x^2 + y^2$$
$$d^2 = t^2 + x^2 - y^2$$

(5.4)

In 3-dimensional space, there are three possible pairs of variables (2-dimensional planes).

If the 3-dimensional distance function is:

$$d^2 = t^2 + x^2 + y^2$$

(5.5)

then the three pairs of variables have the same sign, and so this space has three Euclidean rotational planes but no space-time rotational planes.

If the 3-dimensional distance function is:

$$d^2 = t^2 + x^2 - y^2$$

(5.6)

then only one pair of two variables have the same sign and two pairs of variables have different signs, and so this space has one Euclidean rotational plane but two space-time rotational planes. We have the set of possible 3-dimensional geometric spaces.

$$3 \quad \text{Euclidean rotations}$$
$$1 \quad \text{Euclidean rotations} \quad \& \quad 2 \quad \text{Space-time rotations}$$

(5.7)

In fact, neither of these occurs as an emergent distance function.

In 4-dimensional space, there are six 2-dimensional planes. There are three possible different 4-dimensional distance functions:

---

[8] This is an arithmetic progression.

$$d^2 = t^2 + x^2 + y^2 + z^2$$
$$d^2 = t^2 + x^2 + y^2 - z^2 \tag{5.8}$$
$$d^2 = t^2 + x^2 - y^2 - z^2$$

(Note that $d^2 = t^2 - x^2 - y^2 - z^2$ is equivalent to $d^2 = t^2 + x^2 + y^2 - z^2$.) It is remarkable, and seemingly due to no more than chance, that all three of these possible 4-dimensional distance functions play a role in our universe.

Thus, a 4-dimensional emergent space might be of the forms:

6   Euclidean rotations

3   Euclidean rotations   &   3   Space-time rotations   (5.9)

2   Euclidean rotations   &   4   Space-time rotations

Clearly, our space-time is the 3 Euclidean rotation and 3 space-time rotations one. If it were the case that the A_3[9] emergent distance function was of the form $d^2 = t^2 + x^2 - y^2 - z^2$, then we would have a space-time with 2 Euclidean rotations and 4 space-time rotations. The maths does not give this. However, the quaternion expectation distance function is of the form $d^2 = t^2 + x^2 + y^2 + z^2$, and so we have a quaternion emergent space with 6 Euclidean rotations; there is no time in such a space, and it is not clearly understood how such a timeless space manifests itself. We think the quaternion space is connected to the weak force.

In 5-dimensional space, there are ten 2-dimensional planes. There are three possible different 5-dimensional distance functions:

$$d^2 = t^2 + w^2 + x^2 + y^2 + z^2$$
$$d^2 = t^2 + w^2 + x^2 + y^2 - z^2 \tag{5.10}$$
$$d^2 = t^2 + w^2 + x^2 - y^2 - z^2$$

Thus, a 5-dimensional emergent space might be of the forms:

10   Euclidean rotations

6   Euclidean rotations   &   4   Space-time rotations   (5.11)

4   Euclidean rotations   &   6   Space-time rotations

In 6-dimensional space, there are fifteen 2-dimensional planes. There are four possible different 6-dimensional distance functions:

$$d^2 = t^2 + v^2 + w^2 + x^2 + y^2 + z^2$$
$$d^2 = t^2 + v^2 + w^2 + x^2 + y^2 - z^2$$
$$d^2 = t^2 + v^2 + w^2 + x^2 - y^2 - z^2 \tag{5.12}$$
$$d^2 = t^2 + v^2 + w^2 - x^2 - y^2 - z^2$$

---

[9] See later.

A 6-dimensional emergent space might be of the forms:

$$
\begin{array}{llll}
15 & \text{Euclidean rotations} & & \\
10 & \text{Euclidean rotations} & \& \quad 5 & \text{Space-time rotations} \\
7 & \text{Euclidean rotations} & \& \quad 8 & \text{Space-time rotations} \\
6 & \text{Euclidean rotations} & \& \quad 9 & \text{Space-time rotations}
\end{array}
$$

(5.13)

In 7-dimensional space, there are twenty-one 2-dimensional planes. There are four possible different 7-dimensional distance functions:

$$
\begin{aligned}
d^2 &= t^2 + u^2 + v^2 + w^2 + x^2 + y^2 + z^2 \\
d^2 &= t^2 + u^2 + v^2 + w^2 + x^2 + y^2 - z^2 \\
d^2 &= t^2 + u^2 + v^2 + w^2 + x^2 - y^2 - z^2 \\
d^2 &= t^2 + u^2 + v^2 + w^2 - x^2 - y^2 - z^2
\end{aligned}
$$

(5.14)

Thus, a 7-dimensional emergent space might be of the forms:

$$
\begin{array}{llll}
21 & \text{Euclidean rotations} & & \\
15 & \text{Euclidean rotations} & \& \quad 6 & \text{Space-time rotations} \\
11 & \text{Euclidean rotations} & \& \quad 10 & \text{Space-time rotations} \\
9 & \text{Euclidean rotations} & \& \quad 12 & \text{Space-time rotations}
\end{array}
$$

(5.15)

In 8-dimensional space, there are twenty-eight 2-dimensional planes. There are five possible different 8-dimensional distance functions:

$$
\begin{aligned}
d^2 &= t^2 + s^2 + u^2 + v^2 + w^2 + x^2 + y^2 + z^2 \\
d^2 &= t^2 + s^2 + u^2 + v^2 + w^2 + x^2 + y^2 - z^2 \\
d^2 &= t^2 + s^2 + u^2 + v^2 + w^2 + x^2 - y^2 - z^2 \\
d^2 &= t^2 + s^2 + u^2 + v^2 + w^2 - x^2 - y^2 - z^2 \\
d^2 &= t^2 + s^2 + u^2 + v^2 - w^2 - x^2 - y^2 - z^2
\end{aligned}
$$

(5.16)

Thus, a 8-dimensional emergent space might be of the forms:

$$
\begin{array}{llll}
28 & \text{Euclidean rotations} & & \\
21 & \text{Euclidean rotations} & \& \quad 7 & \text{Space-time rotations} \\
16 & \text{Euclidean rotations} & \& \quad 12 & \text{Space-time rotations} \\
13 & \text{Euclidean rotations} & \& \quad 15 & \text{Space-time rotations} \\
12 & \text{Euclidean rotations} & \& \quad 16 & \text{Space-time rotations}
\end{array}
$$

(5.17)

So, in the case of geometric spaces which support 2-dimensional rotations, we now know what form of emergent distance function we are looking for all dimensions less than or equal to eight. The question is: do any of these distance functions emerge from any of the finite groups?

Chapter 6

The Lower Order Groups

Order 1:

There is only one finite group of order 1. It is the trivial cyclic group C_1. This group is commutative, of course. C_1 has Standard form Cayley table:

$$[a] \tag{6.1}$$

The group C_1 contains only one division algebra. The algebraic matrix form of this division algebra is:

$$[a] \tag{6.2}$$

This division algebra is the real numbers, $\mathbb{R}$. The distance function, norm, of this algebra is the determinant of the algebraic matrix form; this distance function is:

$$d = a \tag{6.3}$$

The polar form of this algebraic matrix form is:

$$\exp([a] + (0)) = \exp([a])\exp([0]) = [e^a][1] \tag{6.4}$$

The e^a is the radial variable (in 1-dimension, this is along the real axis) and the $[1]$ is the rotation matrix containing the 1-dimensional trigonometric function, 1.

The emergent distance function of C_1 is the sum of the distance functions of the isomorphic algebras (only one in this case):

$$d = a \tag{6.5}$$

Order 2:

There is only one finite group of order two. It is the cyclic group C_2. This group is commutative. C_2 has Standard form Cayley table:

$$C_2 \sim \begin{bmatrix} a & b \\ b & a \end{bmatrix} \tag{6.6}$$

Note that the group C_2 can be represented as the two permutation matrices:

$$\begin{bmatrix} 1 & 0 \\ 0 & 1 \end{bmatrix} \quad \& \quad \begin{bmatrix} 0 & 1 \\ 1 & 0 \end{bmatrix} \tag{6.7}$$

This is how one separates the Standard form Cayley table into the individual elements of the finite group. The set of permutation matrices, with matrix multiplication for the group operation, holds all the relations between the group elements.

The group C_2 contains two division algebras. We insert parameters into the algebraic matrix form.

$$\begin{bmatrix} P_{1,1}a & P_{1,2}b \\ P_{2,1}b & P_{2,2}a \end{bmatrix} \tag{6.8}$$

We see that the effect of the parameters in the top row is no more than to scale the variables $\{a,b\}$. We can remove them:

$$\begin{bmatrix} a & b \\ \dfrac{P_{2,1}}{P_{1,2}}b & \dfrac{P_{2,2}}{P_{1,1}}a \end{bmatrix} \equiv \begin{bmatrix} a & b \\ Q_{2,1}b & Q_{2,2}a \end{bmatrix} \tag{6.9}$$

The division algebra must have a multiplicative identity, and so all the parameters on the leading diagonal must be equal:

$$\begin{bmatrix} a & b \\ Q_{2,1}b & a \end{bmatrix} \tag{6.10}$$

Putting $Q_{2,1} = +1$ gives one division algebra. Putting $Q_{2,1} = -1$ gives another division algebra. In fact, the parameters can be set to any absolute value other than zero. Different parameters do not need to be set to the same value. It is the sign of the parameters that separates the division algebras.

In the case of the finite group C_2, these division algebras are the hyperbolic complex numbers, $\mathbb{S}$, and the Euclidean complex numbers, $\mathbb{C}$. The distance functions (norms) of these division algebras are the determinants of the respective algebraic matrix forms. These division algebras and associated distance functions are:

$$\mathbb{S} = \exp\left(\begin{bmatrix} a & b \\ b & a \end{bmatrix}\right) = \begin{bmatrix} e^a & 0 \\ 0 & e^a \end{bmatrix}\begin{bmatrix} \cosh b & \sinh b \\ \sinh b & \cosh b \end{bmatrix}$$
$$d^2 = \det\left(\begin{bmatrix} a & b \\ b & a \end{bmatrix}\right) = a^2 - b^2 \tag{6.11}$$

$$\mathbb{C} = \exp\left(\begin{bmatrix} a & b \\ -b & a \end{bmatrix}\right) = \begin{bmatrix} e^a & 0 \\ 0 & e^a \end{bmatrix}\begin{bmatrix} \cos b & \sin b \\ -\sin b & \cos b \end{bmatrix}$$
$$d^2 = a^2 + b^2 \tag{6.12}$$

The trigonometric functions of these algebras are the well-known 2-dimensional trigonometric functions.

In the order 2 case, as in the order 1 case, there is only one copy of each type of division algebra within the finite group. Therefore the emergent distance functions of these algebras, formed by summing the distance functions of the isomorphic division algebras, are the same as the actual algebras.

The emergent distance functions of C_2 are:

$$d^2 = t^2 - z^2$$
$$d^2 = x^2 + y^2$$

(6.13)

Order 3:

There is only one finite group of order three. It is the cyclic group C_3. This group is commutative. C_3 has Standard form Cayley table:

$$C_3 \sim \begin{bmatrix} a & b & c \\ c & a & b \\ b & c & a \end{bmatrix}$$

(6.14)

This can be separated into the C_3 set of permutation matrices as:

$$\begin{bmatrix} 1 & 0 & 0 \\ 0 & 1 & 0 \\ 0 & 0 & 1 \end{bmatrix} \& \begin{bmatrix} 0 & 1 & 0 \\ 0 & 0 & 1 \\ 1 & 0 & 0 \end{bmatrix} \& \begin{bmatrix} 0 & 0 & 1 \\ 1 & 0 & 0 \\ 0 & 1 & 0 \end{bmatrix}$$

(6.15)

Examples of the relations between these elements are:

$$\begin{bmatrix} 0 & 1 & 0 \\ 0 & 0 & 1 \\ 1 & 0 & 0 \end{bmatrix}\begin{bmatrix} 0 & 0 & 1 \\ 1 & 0 & 0 \\ 0 & 1 & 0 \end{bmatrix} = \begin{bmatrix} 1 & 0 & 0 \\ 0 & 1 & 0 \\ 0 & 0 & 1 \end{bmatrix} \quad \& \quad \begin{bmatrix} 0 & 0 & 1 \\ 1 & 0 & 0 \\ 0 & 1 & 0 \end{bmatrix}\begin{bmatrix} 0 & 0 & 1 \\ 1 & 0 & 0 \\ 0 & 1 & 0 \end{bmatrix} = \begin{bmatrix} 0 & 1 & 0 \\ 0 & 0 & 1 \\ 1 & 0 & 0 \end{bmatrix}$$

(6.16)

The group C_3 contains four division algebras. We insert parameters into the algebraic matrix form. As explained above, we do not bother with the parameters on the top row or on the leading diagonal:

$$\begin{bmatrix} a & b & c \\ P_{2,1}c & a & P_{2,3}b \\ P_{3,1}b & P_{3,2}c & a \end{bmatrix}$$

(6.17)

Multiplying two such matrices together gives:

$$\begin{bmatrix} a & b & c \\ P_{2,1}c & a & P_{2,3}b \\ P_{3,1}b & P_{3,2}c & a \end{bmatrix}\begin{bmatrix} e & f & g \\ P_{2,1}g & e & P_{2,3}f \\ P_{3,1}f & P_{3,2}g & e \end{bmatrix}$$
$$= \begin{bmatrix} ae + P_{2,1}bg + P_{3,1}cf & \sim & \sim \\ \sim & ae + P_{2,3}P_{3,2}bg + P_{2,1}cf & \sim \\ \sim & \sim & ae + P_{3,1}bg + P_{2,3}P_{3,2}cf \end{bmatrix}$$

(6.18)

We know that the elements on the leading diagonal must be equal to form the multiplicative identity of the division algebra and to maintain multiplicative closure of matrix form. Putting:

$$P_{3,2} = \frac{P_{2,1}}{P_{2,3}} \tag{6.19}$$

Leads to:

$$
\begin{bmatrix}
a & b & c \\
P_{2,1}c & a & P_{2,3}b \\
P_{3,1}b & \dfrac{P_{2,1}}{P_{2,3}}c & a
\end{bmatrix}
\begin{bmatrix}
e & f & g \\
P_{2,1}g & e & P_{2,3}f \\
P_{3,1}f & \dfrac{P_{2,1}}{P_{2,3}}g & e
\end{bmatrix}
$$

$$
=
\begin{bmatrix}
ae + P_{2,1}bg + P_{3,1}cf & \sim & \sim \\
\sim & ae + P_{2,1}bg + P_{2,1}cf & \sim \\
\sim & \sim & ae + P_{3,1}bg + P_{2,1}cf
\end{bmatrix} \tag{6.20}
$$

Putting:

$$P_{3,1} = P_{2,1} \tag{6.21}$$

Leads to:

$$
\begin{bmatrix}
a & b & c \\
P_{2,1}c & a & P_{2,3}b \\
P_{2,1}b & \dfrac{P_{2,1}}{P_{2,3}}c & a
\end{bmatrix}
\begin{bmatrix}
e & f & g \\
P_{2,1}g & e & P_{2,3}f \\
P_{2,1}f & \dfrac{P_{2,1}}{P_{2,3}}g & e
\end{bmatrix}
$$

$$
=
\begin{bmatrix}
ae + P_{2,1}bg + P_{2,1}cf & af + be + \dfrac{P_{2,1}}{P_{2,3}}cg & ag + P_{2,3}bf + ce \\[2mm]
P_{2,1}\left(ag + P_{2,3}bf + ce\right) & ae + P_{2,1}bg + P_{2,1}cf & P_{2,3}\left(af + be + \dfrac{P_{2,1}}{P_{2,3}}cg\right) \\[2mm]
P_{2,1}\left(af + be + \dfrac{P_{2,1}}{P_{2,3}}cg\right) & \dfrac{P_{2,1}}{P_{2,3}}\left(ag + P_{2,3}bf + ce\right) & ae + P_{2,1}bg + P_{2,1}cf
\end{bmatrix} \tag{6.22}
$$

We see that the form of the matrix is maintained under multiplication – we have multiplicative closure; of course, the order of the multiplication does not affect the form of the product.

In the case of C_3, we have achieved elimination of unwanted parameters using only the leading diagonal.

In the cases of the higher order groups, we need to consider off diagonal elements also. The calculation can be cumbersome and error prone; it is best done by computer, but, with many groups, we do not know how to program the computer to eliminate the unwanted parameters.

Note that the order of elimination is of no mathematical relevance and that the choice of which parameters to eliminate is of no mathematical relevance. None-the-less, we always eliminate the parameters of a particular type of group in the same way. We do this to try and understand what we are doing; such standard elimination procedure gives clarity rather than fog, we think.

The algebraic matrix form with parameters of the group C_3 is:

$$C_3 \rightarrow \begin{bmatrix} a & b & c \\ P_{2,1}c & a & P_{2,3}b \\ P_{2,1}b & \dfrac{P_{2,1}}{P_{2,3}}c & a \end{bmatrix} \tag{6.23}$$

This matrix, (6.23), is commutative, and so all the algebras which it represents are commutative.

By setting these two parameters to the permutations:

$$\{P_{2,1}=+1 \ \& \ P_{2,3}=+1\}, \ \{P_{2,1}=+1 \ \& \ P_{2,3}=-1\}, \ \{P_{2,1}=-1 \ \& \ P_{2,3}=+1\}, \ \{P_{2,1}=-1 \ \& \ P_{2,3}=-1\} \tag{6.24}$$

we get the four C_3 division algebras:

$$C_3^{++} = \exp\left(\begin{bmatrix} a & b & c \\ c & a & b \\ b & c & a \end{bmatrix}\right) \qquad C_3^{+-} = \exp\left(\begin{bmatrix} a & b & c \\ c & a & -b \\ b & -c & a \end{bmatrix}\right)$$

$$C_3^{-+} = \exp\left(\begin{bmatrix} a & b & c \\ -c & a & b \\ -b & -c & a \end{bmatrix}\right) \qquad C_3^{--} = \exp\left(\begin{bmatrix} a & b & c \\ -c & a & -b \\ -b & c & a \end{bmatrix}\right) \tag{6.25}$$

We give an example:

$$\exp\left(\begin{bmatrix} a & b & c \\ c & a & b \\ b & c & a \end{bmatrix}\right) = \begin{bmatrix} e^a & 0 & 0 \\ 0 & e^a & 0 \\ 0 & 0 & e^a \end{bmatrix}\begin{bmatrix} v_A & v_B & v_C \\ v_C & v_A & v_B \\ v_B & v_C & v_A \end{bmatrix} \tag{6.26}$$

The rightmost of these matrices is the rotation matrix of the spinor space associated (as the Euclidean complex plane is associated with the Euclidean complex numbers) with this division algebra. The functions inside this rotation matrix, $\{v_A,v_B,v_C\}$, the nu-functions[10], are the trigonometric functions of this division algebra space. They are[11]:

[10] They have been known for only a few years, and so it is appropriate to call them new functions.
[11] See: Dennis Morris: Complex Numbers The Higher Dimensional Forms

$$v_A = \frac{1}{3}\left(e^{(b+c)} + 2e^{-\left(\frac{b+c}{2}\right)}\cos\left(\frac{\sqrt{3}}{2}(b-c)\right)\right)$$

$$v_B = \frac{1}{3}\left(e^{(b+c)} + e^{-\left(\frac{b+c}{2}\right)}\left(\sqrt{3}\sin\left(\frac{\sqrt{3}}{2}(b-c)\right) - \cos\left(\frac{\sqrt{3}}{2}(b-c)\right)\right)\right)$$

$$v_C = \frac{1}{3}\left(e^{(b+c)} - e^{-\left(\frac{b+c}{2}\right)}\left(\sqrt{3}\sin\left(\frac{\sqrt{3}}{2}(b-c)\right) + \cos\left(\frac{\sqrt{3}}{2}(b-c)\right)\right)\right)$$

(6.27)

In a way similar to the Euclidean complex numbers having the square root of minus unity within the algebra, so the C_3 division algebras have the cube roots of plus & minus unity as:

$$C_3^{++} \sim a + b\sqrt[3]{+1} + c\sqrt[3]{+1}$$
$$C_3^{+-} \sim a + b\sqrt[3]{-1} + c\sqrt[3]{-1}$$
$$C_3^{-+} \sim a + b\sqrt[3]{-1} + c\sqrt[3]{+1}$$
$$C_3^{--} \sim a + b\sqrt[3]{+1} + c\sqrt[3]{-1}$$

(6.28)

We see that the C_3^{-+} & C_3^{--} algebras are algebraically isomorphic (they have the same roots of plus and minus unity and they are both commutative). We therefore have three distinct non-isomorphic algebras but four separate algebras.

The distance functions (norms) of these algebras are:

$$C_3^{++} \sim d^3 = a^3 + b^3 + c^3 - 3abc$$
$$C_3^{+-} \sim d^3 = a^3 - b^3 - c^3 - 3abc$$
$$C_3^{-+} \sim d^3 = a^3 - b^3 + c^3 - 3abc$$
$$C_3^{--} \sim d^3 = a^3 + b^3 - c^3 - 3abc$$

(6.29)

Adding the distance functions of the isomorphic algebras gives:

$$C_3^{++} \sim d^3 = a^3 + b^3 + c^3 - 3abc$$
$$C_3^{+-} \sim d^3 = a^3 - b^3 - c^3 - 3abc$$

(6.30)

$$C_3^{-+} + C_3^{--} \sim sum\left\{\begin{matrix}d^3 = a^3 - b^3 + c^3 - 3abc\\ d^3 = a^3 + b^3 - c^3 - 3abc\end{matrix}\right\} \sim d^3 = a^3 - 3abc$$

We see that the emergent distance function of the two isomorphic algebras, $d^3 = a^3 - 3abc$, will not reduce to quadratic form and does not support any form of rotation. By this we mean that there are no types of rotation which leave invariant the distance, as measured by this emergent distance function, from the origin in the emergent space. There are no rotations in this emergent space.

Other versions of the Cayley table can be constructed by swapping rows or swapping columns of the Standard form Cayley table:

$$\begin{bmatrix} a & b & c \\ c & a & b \\ b & c & a \end{bmatrix} \rightarrow \begin{bmatrix} b & a & c \\ a & c & b \\ c & b & a \end{bmatrix} \rightarrow \begin{bmatrix} a & c & b \\ b & a & c \\ c & b & a \end{bmatrix} \tag{6.31}$$

We see that the right-most of these, (6.31), is the Standard form Cayley table in which the $\{b,c\}$ variables have been swapped.

Order 4:

There are two finite groups of order 4. It is the cyclic group C_4 and the group $C_2 \times C_2$.

The C_4 group:

The C_4 group is a commutative (abelian) group. The C_4 group has one C_2 subgroup formed from one of the non-identity elements and the identity. The C_4 group has Standard form Cayley table:

$$C_{4.1} \sim \begin{bmatrix} a & b & c & d \\ d & a & b & c \\ c & d & a & b \\ b & c & d & a \end{bmatrix} \tag{6.32}$$

The reader might notice the group elements all run diagonally parallel to the identity on the leading diagonal. We have the products:

$$\begin{bmatrix} 0&1&0&0 \\ 0&0&1&0 \\ 0&0&0&1 \\ 1&0&0&0 \end{bmatrix}\begin{bmatrix} 0&1&0&0 \\ 0&0&1&0 \\ 0&0&0&1 \\ 1&0&0&0 \end{bmatrix} = \begin{bmatrix} 0&0&1&0 \\ 0&0&0&1 \\ 1&0&0&0 \\ 0&1&0&0 \end{bmatrix} \quad \& \quad \begin{bmatrix} 0&1&0&0 \\ 0&0&1&0 \\ 0&0&0&1 \\ 1&0&0&0 \end{bmatrix}\begin{bmatrix} 0&0&1&0 \\ 0&0&0&1 \\ 1&0&0&0 \\ 0&1&0&0 \end{bmatrix} = \begin{bmatrix} 0&0&0&1 \\ 1&0&0&0 \\ 0&1&0&0 \\ 0&0&1&0 \end{bmatrix}$$
$$(6.33)$$

We see the 1's in the permutation matrices 'dance' across the matrices. This is the case for all cyclic finite groups – see C_2 & C_3 above. We can always construct the Standard form Cayley table of a cyclic group by simply writing the variables in parallel diagonal lines across the matrix.

The Standard form Cayley table is not unique for the group C_4. We have two other forms of the Standard form Cayley table:

$$C_{4.2} \sim \begin{bmatrix} a & b & c & d \\ b & a & d & c \\ d & c & a & b \\ c & d & b & a \end{bmatrix} \quad \& \quad C_{4.3} \sim \begin{bmatrix} a & b & c & d \\ c & a & d & b \\ b & d & a & c \\ d & c & b & a \end{bmatrix} \tag{6.34}$$

These Standard form Cayley tables differ from each other by swapping of variables. In each case, a different variable, together with the identity, forms the C_2 subgroup.

The algebraic matrix forms with parameters of the group C_4 is:

$$C_{4.1} \sim \begin{bmatrix} a & b & c & d \\ P_{2,1}d & a & P_{2,3}b & P_{2,4}c \\ \dfrac{P_{2,1}P_{2,4}}{P_{2,3}}c & \dfrac{P_{2,1}}{P_{2,3}}d & a & P_{2,4}b \\ P_{2,1}b & \dfrac{P_{2,1}}{P_{2,3}}c & \dfrac{P_{2,1}}{P_{2,4}}d & a \end{bmatrix} \qquad C_{4.2} \sim \begin{bmatrix} a & b & c & d \\ P_{2,1}b & a & P_{2,3}d & \dfrac{P_{2,1}}{P_{2,3}}c \\ P_{3,1}d & \dfrac{P_{3,1}}{P_{2,3}}c & a & \dfrac{P_{2,1}}{P_{2,3}}b \\ P_{3,1}c & \dfrac{P_{2,3}P_{3,1}}{P_{2,1}}d & P_{2,3}b & a \end{bmatrix} \qquad (6.35)$$

$$C_{4.3} \sim \begin{bmatrix} a & b & c & d \\ P_{2,1}c & a & P_{2,3}d & P_{2,4}b \\ P_{2,1}b & \dfrac{P_{2,1}}{P_{2,4}}d & a & \dfrac{P_{2,1}}{P_{2,3}}c \\ \dfrac{P_{2,1}P_{2,3}}{P_{2,4}}d & \dfrac{P_{2,1}}{P_{2,4}}c & b & a \end{bmatrix}$$

By substitution, we can change these. If, with an eye on the 1st variable of the 3rd row of $C_{4.1}$, we put:

$$P_{2,4} = \frac{P_{3,1}P_{2,3}}{P_{2,1}} \qquad (6.36)$$

Then we get:

$$C_4 \sim \begin{bmatrix} a & b & c & d \\ P_{2,1}d & a & P_{2,3}b & \dfrac{P_{2,3}P_{3,1}}{P_{2,1}}c \\ P_{3,1}c & \dfrac{P_{2,1}}{P_{2,3}}d & a & \dfrac{P_{2,3}P_{3,1}}{P_{2,1}}b \\ P_{2,1}b & \dfrac{P_{2,1}}{P_{2,3}}c & \dfrac{\left(P_{2,1}\right)^2}{P_{2,3}P_{3,1}}d & a \end{bmatrix} \qquad (6.37)$$

If the reader looks ahead to the Standard form Cayley tables of the higher order cyclic groups and considers the left-most column, the reader might conclude that (6.37) is a more consistent way to form the algebraic matrix form of C_4. It is usually presented as in (6.35) for only historical reasons[12].

---

[12] It was your author who made this history when he first discovered these algebras. Unfortunately, your author cannot change history.

These matrices, (6.35), are commutative, and all the algebras which they represent are commutative.

Setting these three parameters, $\{P_{2,1}, P_{2,3}, P_{3,1}\}$, to the different permutations of $\{\pm 1, \pm 1, \pm 1\}$ leads, by taking the exponential of the algebraic matrix form, in each of the three cases, to eight separate C_4 division algebras. However, in each case, of these eight algebras, there are two sets of four isomorphic algebras giving only two distinct types of non-isomorphic division algebras. We refer to these as the H-type algebra and the E-type algebra[13]. All the algebras of the C_4 group are commutative division algebras (algebraic fields).

The H-type C_4 division algebras have one square root of plus unity and two fourth roots of plus unity. The E-type C_4 division algebras have one square root of minus unity and two fourth roots of minus unity.

The square root is associated with the element which generates the C_2 subgroup. The separate division algebras and distance functions have the form (we need to take the exponentials of these forms to get the division algebras):

$$H \quad C_{4.1}^{+++} \sim a + b\sqrt[4]{+1} + c\sqrt[2]{+1} + d\sqrt[4]{+1}$$
$$d^4 = a^4 - b^4 + c^4 - d^4 - 2a^2c^2 + 2b^2d^2 - 4a^2bd + 4ab^2c + 4acd^2 - 4bc^2d \tag{6.38}$$

$$H \quad C_{4.1}^{+--} \sim a + b\sqrt[4]{+1} + c\sqrt[2]{+1} + d\sqrt[4]{+1}$$
$$d^4 = a^4 - b^4 + c^4 - d^4 - 2a^2c^2 + 2b^2d^2 - 4a^2bd - 4ab^2c - 4acd^2 - 4bc^2d \tag{6.39}$$

$$H \quad C_{4.1}^{-+-} \sim a + b\sqrt[4]{+1} + c\sqrt[2]{+1} + d\sqrt[4]{+1}$$
$$d^4 = a^4 - b^4 + c^4 - d^4 + 2a^2c^2 + 2b^2d^2 + 4a^2bd + 4ab^2c + 4acd^2 + 4bc^2d \tag{6.40}$$

$$H \quad C_{4.1}^{--+} \sim a + b\sqrt[4]{+1} + c\sqrt[2]{+1} + d\sqrt[4]{+1}$$
$$d^4 = a^4 - b^4 + c^4 - d^4 - 2a^2c^2 + 2b^2d^2 + 4a^2bd - 4ab^2c - 4acd^2 + 4bc^2d \tag{6.41}$$

$$E \quad C_{4.1}^{++-} \sim a + b\sqrt[4]{-1} + c\sqrt[2]{-1} + d\sqrt[4]{-1}$$
$$d^4 = a^4 + b^4 + c^4 + d^4 + 2a^2c^2 + 2b^2d^2 - 4a^2bd - 4ab^2c + 4acd^2 + 4bc^2d \tag{6.42}$$

$$E \quad C_{4.1}^{+-+} \sim a + b\sqrt[4]{-1} + c\sqrt[2]{-1} + d\sqrt[4]{-1}$$
$$d^4 = a^4 + b^4 + c^4 + d^4 + 2a^2c^2 + 2b^2d^2 + 4a^2bd - 4ab^2c - 4acd^2 + 4bc^2d \tag{6.43}$$

$$E \quad C_{4.1}^{-++} \sim a + b\sqrt[4]{-1} + c\sqrt[2]{-1} + d\sqrt[4]{-1}$$
$$d^4 = a^4 + b^4 + c^4 + d^4 + 2a^2c^2 + 2b^2d^2 + 4a^2bd - 4ab^2c + 4acd^2 - 4bc^2d \tag{6.44}$$

$$E \quad C_{4.1}^{---} \sim a + b\sqrt[4]{-1} + c\sqrt[2]{-1} + d\sqrt[4]{-1}$$
$$d^4 = a^4 + b^4 + c^4 + d^4 + 2a^2c^2 + 2b^2d^2 + 4a^2bd + 4ab^2c - 4acd^2 - 4bc^2d \tag{6.45}$$

We form the emergent distance function by first reducing the above distance functions to quadratic form. Factoring (6.38) gives:

---

[13] The H stands for hyperbolic. The E stands for Euclidean. The 4-dimensional trigonometric functions of these algebras have a hyperbolic or a Euclidean nature.

$$d^4 = \left((a+c)^2 - (b+d)^2\right)\left((a-c)^2 + (b-d)^2\right) \tag{6.46}$$

This will not reduce to quadratic form, and so the H-type algebras will not produce an emergent distance function. The E-type algebras have distance functions which do not factorise and so will not produce an emergent distance function.

Let us just re-iterate the above. To support a given type of rotation, an emergent geometric space must have an emergent distance function which, when reduced to the same number of variables as the dimension of the rotation, becomes the distance function of the given type of rotation.

To form a geometric space of the same dimension as the emergent distance function, all possible pairs of variables in the case of 2-dimensional rotations, or all possible trios of variables in the case of 3-dimensional rotations, must form the distance function of the given lesser dimensional rotation; even more than this required, but this is essential.

What kind of 4-dimensional emergent distance function would support 2-dimensional rotations between all pairs of variables? A Riemann type of distance function like $d^2 = t^2 - x^2 - y^2 - z^2$ does this. What kind of 4-dimensional emergent distance function would support 3-dimensional rotations between all trios of variables? The emergent distance function (if it existed):

$$d^3 = a^3 + b^3 + c^3 + d^3 - 3abc - 3abd - 3bcd \tag{6.47}$$

would support three 3-dimensional rotations of the type C_3^{++} in (6.29). If such a distance function did emerge as the sum of the distance functions of a set of isomorphic division algebras from an order four group, then we would see a 4-dimensional geometric space containing 3-dimensional rotations.

The other forms of the Standard form Cayley table:
However, there are two other forms of the C_4 standard Cayley table. We consider:

$$C_{4.2} \sim \begin{bmatrix} a & b & c & d \\ P_{2,1}b & a & P_{2,3}d & \frac{P_{2,1}}{P_{2,3}}c \\ P_{3,1}d & \frac{P_{3,1}}{P_{2,3}}c & a & \frac{P_{2,1}}{P_{2,3}}b \\ P_{3,1}c & \frac{P_{2,3}P_{3,1}}{P_{2,1}}d & P_{2,3}b & a \end{bmatrix} \tag{6.48}$$

With all three parameters set to plus unity, we have:

$$H \quad C_{4.2}^{+++} \sim a + b\sqrt[2]{+1} + c\sqrt[4]{+1} + d\sqrt[4]{+1}$$
$$d^4 = a^4 + b^4 - c^4 - d^4 - 2a^2b^2 + 2c^2d^2 - 4a^2cd + 4abc^2 + 4abd^2 - 4b^2cd \tag{6.49}$$

Comparing this to the $C_{4.1}$ equivalent with all three parameters set to plus unity, (6.38), we see that the two distance functions are exactly the same except that the $\{b,c\}$ variables have swapped places. It is generally the case that the distance functions of the $C_{4.2}$ form of the C_4 algebras differ from the $C_{4.1}$ form of the C_4 algebras by the swapping of the $\{b,c\}$ variables. This reflects the fact that these two forms of the algebraic matrix form are just the same algebra written in two different bases.

Consider:

$$C_{4.3} \sim \begin{bmatrix} a & b & c & d \\ P_{2,1}c & a & P_{2,3}d & P_{2,4}b \\ P_{2,1}b & \dfrac{P_{2,1}}{P_{2,4}}d & a & \dfrac{P_{2,1}}{P_{2,3}}c \\ \dfrac{P_{2,1}P_{2,3}}{P_{2,4}}d & \dfrac{P_{2,1}}{P_{2,4}}c & b & a \end{bmatrix} \tag{6.50}$$

With all three parameters set to plus unity, we have:

$$H \quad C_{4.3}^{+++} \sim a + b\sqrt[4]{+1} + c\sqrt[4]{+1} + d\sqrt[2]{+1}$$
$$d^4 = a^4 - b^4 - c^4 + d^4 - 2a^2d^2 + 2b^2c^2 - 4a^2cd + 4ab^2c + 4ac^2d - 4bcd^2 \tag{6.51}$$

Subgroups and geometric spaces – a powerful result:

We have shown that the C_4 group does not give rise to any geometric spaces. The *Reduce then Sum* approach to calculating emergent distance functions relies upon reducing the algebras to quadratic form and then adding. We have seen that distance functions of the algebras of the C_4 group cannot be reduced to quadratic form.

Consider a group of order higher than four which has the C_4 group as a subgroup. Set all variables in the larger group to zero except the four which form the C_4 group subgroup. Since this part of the larger group cannot form an emergent distance function, the larger group cannot contain an geometric space.

This is a powerful result.

The powerful result:

The powerful result says that the very many groups that have the C_4 group as a subgroup do not have geometric spaces within them. Furthermore, the same applies to all other groups which, like the C_4 group, cannot be reduced to quadratic form (or cubic for 3-dimensional rotations etc.). We have just shown that a very large proportion of the finite groups do not hold a geometric space. Since, groups of prime order, other than two in the 2-dimensional case, cannot be reduced to the quadratic form, any larger group that has a subgroup of prime order other than two cannot hold a geometric space.

Of course, this powerful result is not limited to only the C_4 group as a subgroup but includes all groups like C_4 which do not hold a geometric space.

In the case of 2-dimensional rotations, we can eliminate any large group that has an odd order because such a group will have an odd powered distance function which cannot factorise into quadratic form.

There is more to this powerful result; it is known that, if p is a prime divisor of the order of a group, then that group has a subgroup of order p. We are thus left with the only possible groups that contain geometric spaces must be of order 2^n.

In the last few paragraphs, we have shown that the great majority of groups cannot hold geometric spaces. However, there are still a lot of groups which we have not eliminated.

Summary of the C_4 distance functions:

Each separate C_4 algebra supports its own rotation, of course, as does every division algebra. None of the emergent distance functions of either of the two types of C_4 division algebras is such that it will support any form of rotation between all pairs of variables (or all triples of variables). Thus there is no geometric space which emerges from the C_4 group.

The $C_2 \times C_2$ Group

We met the $C_2 \times C_2$ group in the first chapter of this book. We include it here more comprehensively.

The $C_2 \times C_2$ group has three C_2 subgroups. Each subgroup is formed from one of the three non-identity elements and the identity. The $C_2 \times C_2$ group is a commutative group. The $C_2 \times C_2$ group has Standard form Cayley table:

$$C_2 \times C_2 \sim \begin{bmatrix} a & b & c & d \\ b & a & d & c \\ c & d & a & b \\ d & c & b & a \end{bmatrix} \tag{7.1}$$

The reader might notice the group elements are symmetrical across the leading diagonal. There is only one Standard form Cayley table for the group $C_2 \times C_2$.

The $C_2 \times C_2$ group is remarkable in that, even though it is a commutative group, it holds non-commutative division algebras like the quaternions. Elimination of six of the nine parameters requires the solution of five linear equations and one quadratic equation.[14] There are two solutions to the quadratic elimination equation; each solution leads to a set of eight division algebras. One set of eight division algebras, the two A_1 algebras and the six A_2 algebras are commutative division algebras. The other set of eight division algebras, the six A_3 algebras and the two quaternion algebras (the quaternions and the anti-quaternions) are non-commutative division algebras.

The two solutions of the quadratic elimination equation gives two algebraic matrix forms with parameters for the group $C_2 \times C_2$:

$$C_2 \times C_2^{Commutative} \sim \begin{bmatrix} a & b & c & d \\ P_{2,1}b & a & P_{2,3}d & \dfrac{P_{2,1}}{P_{2,3}}c \\ P_{3,1}c & \dfrac{P_{2,3}P_{3,1}}{P_{2,1}}d & a & \dfrac{P_{2,1}}{P_{2,3}}b \\ \dfrac{(P_{2,3})^2 P_{3,1}}{P_{2,1}}d & \dfrac{P_{2,3}P_{3,1}}{P_{2,1}}c & P_{2,3}b & a \end{bmatrix} \tag{7.2}$$

---

[14] See: Dennis Morris : The Physics of Empty Space : circa pg. 77

$$C_2 \times C_2^{\,Non\text{-}Commutative} \sim \begin{bmatrix} a & b & c & d \\[6pt] P_{2,1}b & a & P_{2,3}d & \dfrac{P_{2,1}}{P_{2,3}}c \\[10pt] P_{3,1}c & -\dfrac{P_{2,3}P_{3,1}}{P_{2,1}}d & a & -\dfrac{P_{2,1}}{P_{2,3}}b \\[12pt] -\dfrac{(P_{2,3})^2 P_{3,1}}{P_{2,1}}d & \dfrac{P_{2,3}P_{3,1}}{P_{2,1}}c & -P_{2,3}b & a \end{bmatrix} \tag{7.3}$$

The only difference between the two, (7.2) & (7.3), is the presence of the minus signs in (7.3). The minus signs came from choosing the negative root of the quadratic elimination equation.

Both (7.2) and (7.3) are multiplicatively closed.

The commutative $C_2 \times C_2$ algebras:

There are two A_1 commutative $C_2 \times C_2$ algebras in which all three variables are symmetric across the leading diagonal. These algebras are of the form (subject to exponentiation):

$$a + b\sqrt{+1} + c\sqrt{+1} + d\sqrt{+1} \tag{7.4}$$

They have distance functions of the form:

$$\begin{aligned} d^4 &= (a+b+c+d)(a+b-c-d)(a-b+c-d)(a-b-c+d) \\ &= \big((a+b)^2 - (c+d)^2\big)\big((a-b)^2 - (c-d)^2\big) \end{aligned} \tag{7.5}$$

These will not reduce to quadratic form and so cannot produce a geometric space with 2-dimensional rotations.

There are six A_2 commutative $C_2 \times C_2$ algebras in which one variable is symmetric across the leading diagonal and two variables are anti-symmetric across the leading diagonal. These algebras are of the form:

$$a + b\sqrt{+1} + c\sqrt{-1} + d\sqrt{-1} \tag{7.6}$$

They have distance functions of the form:

$$d^4 = \big((a-b)^2 + (c-d)^2\big)\big((a+b)^2 + (c+d)^2\big) \tag{7.7}$$

These will not reduce to quadratic form and so cannot produce a geometric space with 2-dimensional rotations.

Summary of the commutative $C_2 \times C_2$ algebras:

There are no geometric spaces that derive from the commutative $C_2 \times C_2$ division algebras.

The non-commutative $C_2 \times C_2$ algebras:

The non-commutative $C_2 \times C_2$ algebras are the division algebra forms of the 4-dimensional Clifford algebras.

There are two quaternion non-commutative $C_2 \times C_2$ algebras in which all three variables are anti-symmetric across the leading diagonal. These algebras are of the form:

$$a + b\sqrt{-1} + c\sqrt{-1} + d\sqrt{-1} \qquad (7.8)$$

They have distance functions of the form:

$$d^4 = \left(a^2 + b^2 + c^2 + d^2\right)^2$$
$$d^2 = a^2 + b^2 + c^2 + d^2 \qquad (7.9)$$

There are six A_3 non-commutative $C_2 \times C_2$ algebras in which two variables are symmetric across the leading diagonal and one variable is anti-symmetric across the leading diagonal. These algebras are of the form:

$$a + b\sqrt{+1} + c\sqrt{+1} + d\sqrt{-1} \qquad (7.10)$$

They have distance functions of the form:

$$d^4 = \left(a^2 + b^2 - c^2 - d^2\right)^2$$
$$d^2 = a^2 + b^2 - c^2 - d^2 \qquad (7.11)$$

We form the quaternion emergent distance function as the sum of the two reduced quaternion distance functions:

$$sum \begin{cases} d^2 = a^2 + b^2 + c^2 + d^2 \\ d^2 = a^2 + b^2 + c^2 + d^2 \end{cases} \Rightarrow \mathbb{H}^{EDF} \sim d^2 = a^2 + b^2 + c^2 + d^2 \qquad (7.12)$$

We form the A_3 emergent distance function as the sum of the six reduced A_3 distance functions:

$$sum \begin{cases} d^2 = a^2 + b^2 - c^2 - d^2 \\ d^2 = a^2 + b^2 - c^2 - d^2 \\ d^2 = a^2 - b^2 + c^2 - d^2 \\ d^2 = a^2 - b^2 + c^2 - d^2 \\ d^2 = a^2 - b^2 - c^2 + d^2 \\ d^2 = a^2 - b^2 - c^2 + d^2 \end{cases} \Rightarrow A_3{}^{EDF} \sim d^2 = 3a^2 - b^2 - c^2 - d^2 \qquad (7.13)$$

The emergent distance functions of the two types of $C_2 \times C_2$ division algebras are:

$$\mathbb{H}^{EDF} \sim d^2 = a^2 + b^2 + c^2 + d^2 \qquad (7.14)$$

$$A_3{}^{EDF} \sim d^2 = 3a^2 - b^2 - c^2 - d^2 \qquad (7.15)$$

We see that the A_3 emergent distance function is of the form of our 4-dimensional space-time. The 3 in the A_3 emergent distance function can be dispensed with because we know that the a variable is time and we can choose units to measure time that will dispense with the 3.

Setting any two variables to zero in the quaternion emergent distance function gives the distance functions of the 2-dimensional Euclidean rotations in the C_2 group. The quaternion emergent distance function therefore supports 2-dimensional Euclidean rotations in six 2-dimensional planes. The quaternion emergent space is a geometric space.

It might seem that the quaternion emergent distance function also supports 4-dimensional quaternion rotation. Emergent spaces have real variables; as such the reader might wonder whether or not the quaternion emergent space can support a spinor rotation like quaternion rotation. The 2-dimensional rotations are all spinor rotations. Clearly, our 4-dimensional space-time has real variables and supports the 2-dimensional spinor rotations. It would seem that quaternion spinor rotation is supported by the quaternion emergent space. Such quaternion spinor rotation is associated with double cover electron spin; we present the quaternion rotation matrix[15]:

$$\begin{bmatrix} \cos\lambda & \frac{b}{\lambda}\sin\lambda & \frac{c}{\lambda}\sin\lambda & \frac{d}{\lambda}\sin\lambda \\ -\frac{b}{\lambda}\sin\lambda & \cos\lambda & -\frac{d}{\lambda}\sin\lambda & \frac{c}{\lambda}\sin\lambda \\ -\frac{c}{\lambda}\sin\lambda & \frac{d}{\lambda}\sin\lambda & \cos\lambda & -\frac{b}{\lambda}\sin\lambda \\ -\frac{d}{\lambda}\sin\lambda & -\frac{c}{\lambda}\sin\lambda & \frac{b}{\lambda}\sin\lambda & \cos\lambda \end{bmatrix} \quad :\lambda = \sqrt{b^2+c^2+d^2} \quad (7.16)$$

Adjusting the co-ordinate system so that the 4-dimensional quaternion rotation lies in the $\{a,b\}$ plane gives:

$$\begin{bmatrix} \cos\lambda & \frac{b}{\lambda}\sin\lambda & 0 & 0 \\ -\frac{b}{\lambda}\sin\lambda & \cos\lambda & 0 & 0 \\ 0 & 0 & \cos\lambda & -\frac{b}{\lambda}\sin\lambda \\ 0 & 0 & \frac{b}{\lambda}\sin\lambda & \cos\lambda \end{bmatrix} \quad :\lambda = \sqrt{b^2} \quad (7.17)$$

Looking at the positions of the minus signs in the matrix (7.17), we see that we have rotation in both the clockwise direction and the anti-clockwise rotation at the same time. Alternatively, putting both, say, $b = \frac{\pi}{3}$ & $b = -\frac{\pi}{3}$ into the rotation matrix gives the same rotation. We have double cover.

---

[15] See: Dennis Morris: Quaternions

In the same way that the Euclidean complex numbers, $\mathbb{C}$, are the splitting field of the polynomial $x^2 + 1 = 0$, so the quaternions are the splitting field of the polynomial $x^8 - 72x^6 + 180x^4 - 144x^2 + 36 = 0$

We have two geometrical spaces. Most people think we live in only one geometric space, 4-dimensional space-time. We seem to have one geometrical space too many.

Our 4-dimensional space-time does not allow anything to move at the velocity of light. Therefore light is not in our 4-dimensional space-time. It seems that light, and all quantum physics is in the quaternion emergent space. We know that all classical physics is within our 4-dimensional space-time.[16] Observation seems to support the view that we live in two geometric spaces.

Larger groups with $C_2 \times C_2$ as a subgroup:

Having found geometric spaces within the $C_2 \times C_2$ group, we wonder if we will find such geometric spaces in larger groups which have $C_2 \times C_2$ as a subgroup.

Summary:

The quaternion emergent space and the A_3 emergent space are both geometric spaces.

---

[16] See: Dennis Morris *Upon General Relativity*

Chapter 8

Continuous Groups

Continuous groups are each a complete rotational surface. An example is the circle of unit radius; another example is the surface of a sphere in $\mathbb{R}^3$ of unit radius.

There are three types of rotational surfaces.

a) Rotations like the surface of the sphere or the circle; these are called compact continuous groups.
b) Space-time rotations which we call the Lorentz boosts; these are called non-compact continuous groups.
c) The set of points defined by a division algebra rotation matrix such as the quaternion rotation matrix.

Rotational surfaces are defined by the fact that every point of them is distance one unit from the origin as measured by a particular distance function. Different distance functions give different continuous groups, but there are three cases, two 2-dimensional and one 4-dimensional, where two different continuous groups are associated with a particular distance function - see later.

Division algebra continuous groups:

There are an infinite number of continuous groups associated with the division algebras. Examples are the rotational surfaces of the two 2-dimensional division algebras which are the Euclidean complex numbers, $\mathbb{C}$, and the hyperbolic complex numbers, $\mathbb{S}$; these are associated with the distance functions (determinants of the algebraic matrix form) $dist^2 = x^2 + y^2$ and $dist^2 = t^2 - z^2$ respectively. One of these is a circle and the other is a hyperbola. Other examples are the four 3-dimensional rotational surfaces of the four 3-dimensional division algebras which respect distance functions of the form $dist^3 = a^3 + b^3 + c^3 - 3abc$.

Each finite group has within it one or more division algebras. Each type of division algebra has a polar form which includes a rotation matrix. The set of points defined by that rotation matrix is a continuous group that is continuous in as many directions as there are real variables within the rotation matrix, and so we have an infinity of such continuous groups. In a division algebra space of dimension n, we see that these continuous groups are each defined by a single matrix with $(n-1)$ parameters (angles) appearing as arguments of the n-dimensional trigonometric functions within that matrix. Most, but not the quaternion algebras or the Euclidean complex numbers, $\mathbb{C}$, of these continuous groups are non-compact. These continuous groups have not yet been classified.

Our interest in these division algebra type of continuous groups is limited to both the 2-dimensional rotation matrices, the two quaternion rotation matrices with distance functions of the form $dist^2 = a^2 + b^2 + c^2 + d^2$ and the six A_3 rotation matrices with distance functions of the form $dist^2 = t^2 + x^2 - y^2 - z^2$.

The continuous groups of the geometric spaces:

There are only two geometric spaces which emerge from the finite group algebras. These two spaces have the distance functions:

$$dist^2 = t^2 - x^2 - y^2 - z^2 \qquad\qquad dist^2 = a^2 + b^2 + c^2 + d^2 \tag{8.1}$$

Note the coincidence of the quaternion distance function. However, since these are the distance functions of geometric spaces, they contain geometric sub-spaces of lesser dimension. Ignoring the trivial 1-dimensional case, these subspaces have the distance functions:

$$dist^2 = t^2 - x^2 - y^2 \qquad\qquad dist^2 = a^2 + b^2 + c^2$$
$$dist^2 = t^2 - x^2 \qquad\qquad dist^2 = a^2 + b^2 \tag{8.2}$$

We therefore have a total of only six continuous groups of this nature. Two of these, (8.2), are coincident with the 2-dimensional division algebra distance functions.

Whereas the division algebra type of continuous groups are defined by a single matrix with $(n-1)$ parameters, the geometric space groups are defined by a number of matrices which each contain a single parameter (angle). The number of matrices is equal to the number of different pairs of variables in the distance function.

We list the elements geometric space continuous groups together with the distance functions which they respect.

a) The two 2-dimensional geometric space continuous groups are each defined by a single matrix:

$$\begin{bmatrix} \cos\theta & \sin\theta \\ -\sin\theta & \cos\theta \end{bmatrix} \qquad\qquad \begin{bmatrix} \cosh\chi & \sinh\chi \\ -\sinh\chi & \cosh\chi \end{bmatrix}$$
$$dist^2 = x^2 + y^2 \qquad\qquad dist^2 = t^2 - z^2 \tag{8.3}$$
$$\{t, x, y, z\} \in \mathbb{R}$$

These two rotation matrices are exactly the same as the two 2-dimensional division algebra continuous group matrices except that in the geometric case both the variables are real variables whereas in the division algebra case only one of the variables is a real variable.

b) The first 3-dimensional geometric space continuous group is defined by the three matrices:

$$\begin{bmatrix} \cosh\chi_1 & \sinh\chi_1 & 0 \\ \sinh\chi_1 & \cosh\chi_1 & 0 \\ 0 & 0 & 1 \end{bmatrix} \quad \begin{bmatrix} \cosh\chi_2 & 0 & \sinh\chi_2 \\ 0 & 1 & 0 \\ \sinh\chi_2 & 0 & \cosh\chi_2 \end{bmatrix} \quad \begin{bmatrix} 1 & 0 & 0 \\ 0 & \cos\theta & \sin\theta \\ 0 & -\sin\theta & \cos\theta \end{bmatrix}$$
$$dist^2 = t^2 - x^2 - y^2 \tag{8.4}$$
$$\{t, x, y\} \in \mathbb{R}$$

The second 3-dimensional geometric space continuous group is defined by the three matrices:

$$\begin{bmatrix} \cos\theta_1 & \sin\theta_1 & 0 \\ -\sin\theta_1 & \cos\theta_1 & 0 \\ 0 & 0 & 1 \end{bmatrix} \quad \begin{bmatrix} \cos\theta_2 & 0 & \sin\theta_2 \\ 0 & 1 & 0 \\ -\sin\theta_2 & 0 & \cos\theta_2 \end{bmatrix} \quad \begin{bmatrix} 1 & 0 & 0 \\ 0 & \cos\theta_3 & \sin\theta_3 \\ 0 & -\sin\theta_3 & \cos\theta_3 \end{bmatrix}$$

$$dist^2 = x^2 + y^2 + z^2 \tag{8.5}$$

$$\{x, y, z\} \in \mathbb{R}$$

In general, pairs of variables connected by a single minus sign and a single plus sign have hyperbolic rotations and pairs of variables connected by two plus signs have Euclidean rotations.

c) The first 4-dimensional geometric space continuous group is defined by the six matrices:

$$\begin{bmatrix} \cosh\chi_1 & \sinh\chi_1 & 0 & 0 \\ \sinh\chi_1 & \cosh\chi_1 & 0 & 0 \\ 0 & 0 & 1 & 0 \\ 0 & 0 & 0 & 1 \end{bmatrix} \quad \begin{bmatrix} \cosh\chi_2 & 0 & \sinh\chi_2 & 0 \\ 0 & 1 & 0 & 0 \\ \sinh\chi_2 & 0 & \cosh\chi_2 & 0 \\ 0 & 0 & 0 & 1 \end{bmatrix} \quad \begin{bmatrix} \cosh\chi_3 & 0 & 0 & \sinh\chi_3 \\ 0 & 1 & 0 & 0 \\ 0 & 0 & 1 & 0 \\ \sinh\chi_3 & 0 & 0 & \cosh\chi_3 \end{bmatrix}$$

$$\begin{bmatrix} 1 & 0 & 0 & 0 \\ 0 & \cos\theta_1 & \sin\theta_1 & 0 \\ 0 & -\sin\theta_1 & \cos\theta_1 & 0 \\ 0 & 0 & 0 & 1 \end{bmatrix} \quad \begin{bmatrix} 1 & 0 & 0 & 0 \\ 0 & \cos\theta_2 & 0 & \sin\theta_2 \\ 0 & 0 & 1 & 0 \\ 0 & -\sin\theta_2 & 0 & \cos\theta_2 \end{bmatrix} \quad \begin{bmatrix} 1 & 0 & 0 & 0 \\ 0 & 1 & 0 & 0 \\ 0 & 0 & \cos\theta_3 & \sin\theta_3 \\ 0 & 0 & -\sin\theta_3 & \cos\theta_3 \end{bmatrix}$$

$$\tag{8.6}$$

$$dist^2 = t^2 - x^2 - y^2 - z^2$$

$$\{t, x, y, z\} \in \mathbb{R}$$

The second 4-dimensional geometric space continuous group is defined by the six matrices:

$$\begin{bmatrix} \cos\theta_1 & \sin\theta_1 & 0 & 0 \\ -\sin\theta_1 & \cos\theta_1 & 0 & 0 \\ 0 & 0 & 1 & 0 \\ 0 & 0 & 0 & 1 \end{bmatrix} \quad \begin{bmatrix} \cos\theta_2 & 0 & \sin\theta_2 & 0 \\ 0 & 1 & 0 & 0 \\ -\sin\theta_2 & 0 & \cos\theta_2 & 0 \\ 0 & 0 & 0 & 1 \end{bmatrix} \quad \begin{bmatrix} \cos\theta_3 & 0 & 0 & \sin\theta_3 \\ 0 & 1 & 0 & 0 \\ 0 & 0 & 1 & 0 \\ -\sin\theta_3 & 0 & 0 & \cos\theta_3 \end{bmatrix}$$

$$\begin{bmatrix} 1 & 0 & 0 & 0 \\ 0 & \cos\theta_4 & \sin\theta_4 & 0 \\ 0 & -\sin\theta_4 & \cos\theta_4 & 0 \\ 0 & 0 & 0 & 1 \end{bmatrix} \quad \begin{bmatrix} 1 & 0 & 0 & 0 \\ 0 & \cos\theta_5 & 0 & \sin\theta_5 \\ 0 & 0 & 1 & 0 \\ 0 & -\sin\theta_5 & 0 & \cos\theta_5 \end{bmatrix} \quad \begin{bmatrix} 1 & 0 & 0 & 0 \\ 0 & 1 & 0 & 0 \\ 0 & 0 & \cos\theta_6 & \sin\theta_6 \\ 0 & 0 & -\sin\theta_6 & \cos\theta_6 \end{bmatrix}$$

$$\tag{8.7}$$

$$dist^2 = a^2 + b^2 + c^2 + d^2$$

$$\{a, b, c, d\} \in \mathbb{R}$$

These six sets of matrices are the entire six continuous groups of the geometric spaces. They hold invariant the distance functions associated with them. For example:

$$\begin{bmatrix} \cosh \chi & \sinh \chi & 0 & 0 \\ \sinh \chi & \cosh \chi & 0 & 0 \\ 0 & 0 & 1 & 0 \\ 0 & 0 & 0 & 1 \end{bmatrix} \begin{bmatrix} t \\ x \\ y \\ z \end{bmatrix} = \begin{bmatrix} t\cosh \chi + x\sinh \chi \\ t\sinh \chi + x\cosh \chi \\ y \\ z \end{bmatrix}$$

$$\left(t\cosh \chi + x\sinh \chi \right)^2 - \left(t\sinh \chi + x\cosh \chi \right)^2 - y^2 - z^2 \qquad (8.8)$$
$$= t^2 \cosh^2 \chi + x^2 \sinh^2 \chi + 2tx\cosh \chi \sinh \chi$$
$$- t^2 \sinh^2 \chi - x^2 \cosh^2 \chi - 2tx\cosh \chi \sinh \chi$$
$$- y^2 - z^2$$
$$= t^2 - x^2 - y^2 - z^2$$

The continuous groups of the division algebras:

We are interested in only ten division algebras. These are the six A_3 algebras, the two quaternion algebras, and the two 2-dimensional algebras. The two 2-dimensional continuous groups are defined by the two rotation matrices:

$$\begin{bmatrix} \cos \theta & \sin \theta \\ -\sin \theta & \cos \theta \end{bmatrix} \qquad\qquad \begin{bmatrix} \cosh \chi & \sinh \chi \\ -\sinh \chi & \cosh \chi \end{bmatrix}$$
$$dist^2 = x^2 + y^2 \qquad\qquad\qquad dist^2 = t^2 - z^2 \qquad (8.9)$$
$$\{t, x\} \in \mathbb{R} \quad : \quad y \propto \sqrt{-1} \quad : \quad z \propto \sqrt{+1}$$

Which match the above (8.3) except for the nature of the variables. The above (8.3) exist only as sub-groups of the 4-dimensional geometric continuous groups. These two continuous groups, (8.9) exist as independent division algebras.

The two quaternion rotation matrices are:

$$\mathbb{H}_{Rot} = \begin{bmatrix} \cos \lambda & \dfrac{b}{\lambda}\sin \lambda & \dfrac{c}{\lambda}\sin \lambda & \dfrac{d}{\lambda}\sin \lambda \\ -\dfrac{b}{\lambda}\sin \lambda & \cos \lambda & -\dfrac{d}{\lambda}\sin \lambda & \dfrac{c}{\lambda}\sin \lambda \\ -\dfrac{c}{\lambda}\sin \lambda & \dfrac{d}{\lambda}\sin \lambda & \cos \lambda & -\dfrac{b}{\lambda}\sin \lambda \\ -\dfrac{d}{\lambda}\sin \lambda & -\dfrac{c}{\lambda}\sin \lambda & \dfrac{b}{\lambda}\sin \lambda & \cos \lambda \end{bmatrix} \qquad \mathbb{H}_{Rot-Anti} = \begin{bmatrix} \cos \lambda & \dfrac{b}{\lambda}\sin \lambda & \dfrac{c}{\lambda}\sin \lambda & \dfrac{d}{\lambda}\sin \lambda \\ -\dfrac{b}{\lambda}\sin \lambda & \cos \lambda & \dfrac{d}{\lambda}\sin \lambda & -\dfrac{c}{\lambda}\sin \lambda \\ -\dfrac{c}{\lambda}\sin \lambda & -\dfrac{d}{\lambda}\sin \lambda & \cos \lambda & \dfrac{b}{\lambda}\sin \lambda \\ -\dfrac{d}{\lambda}\sin \lambda & \dfrac{c}{\lambda}\sin \lambda & -\dfrac{b}{\lambda}\sin \lambda & \cos \lambda \end{bmatrix}$$

$$(8.10)$$

$$\lambda = \sqrt{b^2 + c^2 + d^2}$$

$$dist^2 = a^2 + b^2 + c^2 + d^2 \qquad\qquad (8.11)$$

$$a \in \mathbb{R} \quad : \quad b \propto \sqrt{-1} \quad : \quad c \propto \sqrt{-1} \quad : \quad d \propto \sqrt{-1}$$

The six matrices above, (8.7), hold invariant the same distance function as the two quaternion algebras. We see that we have two different types of continuous group, with different rotational surfaces, holding the same distance function invariant. Whereas the four variables of (8.7) are real variables, within a quaternion, only one variable is real.

There is a significant difference between the 4-dimensional continuous group defined by the six matrices, (8.7), and the 4-dimensional continuous group defined by the quaternion rotation matrix, (8.10). There are six continuous real parameters defining the geometric continuous group, (8.7). There are only three continuous real parameters defining the quaternion continuous group, (8.10). The quaternion continuous group is continuous in only three 2-dimensional planes. This is reminiscent of electron spin.

Hang on! I hear the reader cry. The quaternion rotation matrix is rotation in a 4-dimensional space. Our experience of 4-dimensional space-time leads us to understand that there are six 2-dimensional planes in which we can continuously rotate in 4-dimensional space (six pairs of variables). Rotation in each of the six 2-dimensional planes is specified by one of the six parameters in the six matrices (one in each) given in (8.7). Since we have only three parameters in the quaternion rotation matrix, we can rotate continuously in only three 2-dimensional planes. Weird stuff quaternion space. It's even weirder. The '2-dimensional' rotations in quaternion space are actually 4-dimensional 2-dimensional rotations – they are double cover rotations.

The wave-function of quantum physics:
The physically observable properties of a quantum physics wave-function are in the form of the modulus of the wave-function. If the wave-function is taken to be a single complex number, $\mathbb{C}$, as in the Schrödinger equation, that modulus is of the form $|\psi|^2 = x^2 + y^2$; if the wave-function is taken to be a pair of complex numbers, $\mathbb{C}^2$, a Weyl spinor, as in the Pauli-Schrödinger equation, that modulus is of the form $|\psi|^2 = a^2 + b^2 + c^2 + d^2$. The continuous groups which hold the modulus of the wave-function invariant are the Euclidean geometric rotations and the quaternion rotations. These are called unitary continuous groups.

Within our 4-dimensional space-time:
Our 4-dimensional space-time is a geometric space. Thus any physics within our 4-dimensional space-time must be invariant under rotation with respect to (8.3) & (8.4) & (8.5) & (8.6) & (8.7). If the physics is invariant under rotation with respect to these rotations, then it is 'accidentally' invariant under rotation with respect to (8.9) & (8.10). However, there are big differences between these types of rotations.

Within our 4-dimensional space-time, we can rotate 2-dimensionally in any of the six 2-dimensional planes. In quaternion space, we can rotate 2-dimensionally in only the three planes formed from the real axis and one of the imaginary variables. The quaternion 2-dimensional rotations do not 'hang together' into a geometric space.

Quaternion rotation is not rotation about an axis; quaternion rotation is 4-dimensional rotation – there are no constant eigenvectors of the rotation matrix and so no axis of rotation. We cannot see 4-dimensional rotation in our fabricated 4-dimensional space-time, and so we assume quaternion rotation in our 4-dimensional space-time will be manifest as a 2-dimensional rotation. However, looking at (8.10) , we see there are two 2-dimensional rotations, clockwise and anti-clockwise within the quaternion rotation. More clearly, we can change the quaternion co-ordinate system to give a pseudo 2-dimensional rotation (the position of the minus signs):

$$\mathbb{H}_{Rot} = \begin{bmatrix} \cos\sqrt{b^2} & \sin\sqrt{b^2} & 0 & 0 \\ -\sin\sqrt{b^2} & \cos\sqrt{b^2} & 0 & 0 \\ 0 & 0 & \cos\sqrt{b^2} & -\sin\sqrt{b^2} \\ 0 & 0 & \sin\sqrt{b^2} & \cos\sqrt{b^2} \end{bmatrix} \tag{8.12}$$

We have double rotation.

The A_3 algebras:

A typical A_3 rotation matrix is:

$$A_{3\,Rot} = \begin{bmatrix} \cosh\lambda & \dfrac{b}{\lambda}\sinh\lambda & \dfrac{c}{\lambda}\sinh\lambda & \dfrac{d}{\lambda}\sinh\lambda \\ -\dfrac{b}{\lambda}\sinh\lambda & \cosh\lambda & -\dfrac{d}{\lambda}\sinh\lambda & \dfrac{c}{\lambda}\sinh\lambda \\ \dfrac{c}{\lambda}\sinh\lambda & -\dfrac{d}{\lambda}\sinh\lambda & \cosh\lambda & -\dfrac{b}{\lambda}\sinh\lambda \\ \dfrac{d}{\lambda}\sinh\lambda & \dfrac{c}{\lambda}\sinh\lambda & \dfrac{b}{\lambda}\sinh\lambda & \cosh\lambda \end{bmatrix}$$

$$\lambda = \sqrt{-b^2 + c^2 + d^2} \tag{8.13}$$

$$dist^2 = a^2 + b^2 - c^2 - d^2$$

If we adjust the co-ordinate system such that the variables $b = d = 0$, we have the pseudo 2-dimensional rotation:

$$A_{3\,Rot} = \begin{bmatrix} \cosh\sqrt{c^2} & 0 & \sinh\sqrt{c^2} & 0 \\ 0 & \cosh\sqrt{c^2} & 0 & \sinh\sqrt{c^2} \\ \sinh\sqrt{c^2} & 0 & \cosh\sqrt{c^2} & 0 \\ 0 & \sinh\sqrt{c^2} & 0 & \cosh\sqrt{c^2} \end{bmatrix} \tag{8.14}$$

This is a double cover rotation in space-time - $\sinh(\ )$ is an odd function.

41

If we adjust the co-ordinate system such that the variables $c = d = 0$, we have the pseudo 2-dimensional rotation:

$$A_{3\,Rot} = \begin{bmatrix} \cos\sqrt{b^2} & \sin\sqrt{b^2} & 0 & 0 \\ -\sin\sqrt{b^2} & \cos\sqrt{b^2} & 0 & 0 \\ 0 & 0 & \cos\sqrt{b^2} & -\sin\sqrt{b^2} \\ 0 & 0 & \sin\sqrt{b^2} & \cos\sqrt{b^2} \end{bmatrix} \tag{8.15}$$

This is a double cover rotation in space.

Lie groups:

Readers who are familiar with Lie group theory will know that there are an infinite number of Lie groups which are thought of as being continuous groups in Riemann spaces of higher dimensions. For example, the Lie group $SU(3)$ is a set of eight generators each of which corresponds to a matrix with a real single variable (angle) for an argument. The existence of these higher dimensional Lie groups is based upon the assumption that there are geometric spaces of dimension higher than four which have Riemann type quadratic forms for distance functions and thereby support 2-dimensional rotation between every pair of variables. There are no such geometric spaces within the finite groups; your author contends that such higher dimensional geometric spaces do not exist, and so your author contends that these higher order Lie groups do not exist unless they coincide with a division algebra rotation matrix continuous group in the way that the Lie group $SU(2)$ coincides with the quaternion division algebras. There are no division algebras with eight non-commutative elements, and so $SU(3)$ does not coincide with a division algebra. There are 8-dimensional division algebras (Clifford algebras) which have six non-commutative elements, but there is no Lie group with six generators. The reader will form their own opinion regarding Lie groups.

Preview of next chapter:

The last two chapters have been long and detailed. We have used the lower order groups to demonstrate the concepts involved and the bones of the calculation techniques. We are now equipped to look at the higher order groups. The amount of calculation involved increases exponentially with the order of the group; we must use a computer to deal with these higher order groups. The next chapter gives details of the computer program we use.

Chapter 9

The Computer Program

In this book, from calculations done by computer, we assert that we have proven there are only two geometric spaces which emerge from the finite groups we have examined. Computer proofs are not universally accepted by the mathematics fraternity. The reader might know that the proof of the 4-colour map theorem is by computer and that some mathematicians do not accept it as a valid mathematical proof because it relies upon a computer. This is not some mathematicians being awkward. There is a theorem of logic which proves it is impossible to know with certainty that a computer program, no matter how small, does not contain a bug. We do not really know what is happening inside a computer's processing chip or inside a computer program.

There is a counter argument that says we cannot be certain that even a non-computer mathematical proof is a proof. The greatest mathematician to have ever lived, Carl Gauss, proved the *Theorema Egregium* (king of theorems) in 1827. His proof was widely accepted as a correct proof by the mathematicians of the time. Since then, later mathematicians have found four separate errors in Gauss's proof. The errors have been corrected and the theorem stands, but, if Gauss and ten thousand other mathematicians missed the errors, how can we be sure of any mathematical proof? When all is said and done, a mathematical proof is no more than a piece of mathematics which has been studied by thousands of mathematicians without a single one of them spotting any errors. It is good, but it is not perfect. Against this are proofs such as the proof of the irrational nature of the square roots of numbers ending in 2, 3, 7, & 8 given by your author elsewhere[17] which are 'obviously' correct without question.

Since our proof is by computer, we should present the computer program to the world. There are insights into the nature of the division algebras within the form of the computer program, and so this is not just a dry presentation of code that is difficult to unravel.

Stop press: In January 2016, an error was discovered in the computer program. The error was not major, but it did effect the number of free parameters which the computer calculated for groups in which had free parameters in columns ten and above. The error was in the 'We will collect into a list all the parameters' part of the program listed below. The fourth line of this part of the program reads

if *length*($PARAM[row, col]$) = 9 **then**

it should read

if *length*($PARAM[row, col]$) = 9 **or** *length*($PARAM[row, col]$) = 10 **then**

Yes! computer proofs can be questionable.

---

[17] Dennis Morris: From Where Comes the Universe, chapter three.

Introduction:

We give here the computer program used by your author to search the finite groups for division algebras and geometric spaces based on an emergent distance function.

The computer program is written in the Maple 17 language.

The first part of the computer program:

The first part of the computer program loads the required packages, sets the size of the square matrices to the order of the group, puts in the name of the group, and defines the matrices which we will need in the program.

```
>  restart :
>  with(LinearAlgebra) : with(ListTools) : with(combinat) :
>  SIZE := 6;
        # SIZE is the order of the cyclic group

>  THE_GROUP := "Cyclic order 6" :
        # We make a note of which group we are working with for presentation later

>  ANDY := Matrix(SIZE) : SYD := Matrix(SIZE) :
        # We will use the matrices ANDY & SYD to form the product as we look for multiplicative
        closure.

>  PARAM := Matrix(SIZE) : VARMAT := Matrix(SIZE) :
        # We will use the matrices PARAM & VARMAT to test that we have multiplicative closure
        and other analysis.

>  VARMAT2 := Matrix(SIZE) : VARMAT3 := Matrix(SIZE) :
        # We will need two spare copies of VARMAT.

>  VARMAT4 := Matrix(SIZE) :
```

The matrices ANDY and SYD are to be copies of the algebraic matrix form (variables and parameters) of the group with different variables but the same parameters. The variables within ANDY are to be of the form $A_0, A_1, ... A_{(SIZE-1)}$. The variables within SYD are to be of the form $S_0, S_1, ... S_{(SIZE-1)}$. PARAM is a matrix which will contain only the parameters of the algebraic matrix form of the group (no variables). These parameters will be of the form $P_{row, column}$. The VARMAT matrices will be copies of the variables within the algebraic matrix form; they will be copies of the Standard form Cayley table written with different variables. The variables in VARMAT will be $A_0 .. A_{(SIZE-1)}$, the variables in VARMAT2 will be $X_0 .. X_{(SIZE-1)}$, and the variables in VARMAT3 will be $B_0 .. B_{(SIZE-1)}$ and the variables in VARMAT4 will be $C_0 .. C_{(SIZE-1)}$. There is no significance in the choice of Latin letters used.

The second part of the computer program:

The second part of the computer program fills the matrices defined above with the variables and parameters ready to begin eliminating parameters to form the algebraic matrix form of the group. This part of the program will vary from group to group because the Standard form Cayley tables vary from group to group, but each group will use the same variables and the same names for the matrices. This enables us to use the same standard analysis code for each group later in the program.

For the cyclic groups, the code that sets up the matrices is:

```
>   dance := 0 :
            # We install the variables  and the parameters of the cyclic group

>   for row from 1 to SIZE do
>   dance := dance + 1 :
>   for col from 1 to SIZE do
>   VARMAT[row, col] := A[(col − dance) mod SIZE]
>   VARMAT2[row, col] := X[(col − dance) mod SIZE]
>   VARMAT3[row, col] := B[(col − dance) mod SIZE]
>   VARMAT4[row, col] := C[(col − dance) mod SIZE]
>   ANDY[row, col] := P[row, col]·A[(col − dance) mod SIZE]
>   SYD[row, col] := P[row, col]·S[(col − dance) mod SIZE]
>   PARAM[row, col] := P[row, col] :
>   end do;
>   end do:                       # Matrices are now set up

>   for col from 1 to SIZE do
            # We put the top row parameters to unity and we put the leading diagonal parameters to
            unity

>   P[1, col] := 1;
>   P[col, col] := 1;
>   end do:                       # We are now ready to begin eliminating parameters
>
>   #print(ANDY);                 # If we want to see the result of our work, we print that result.
```

The cyclic groups are amenable to being set up by standard code regardless of order. More often, the matrices of a given group are set up by hand. Using order four cyclic group as an example, we now have:

$$PARAM = \begin{bmatrix} 1 & 1 & 1 & 1 \\ P_{2,1} & 1 & P_{2,3} & P_{2,4} \\ P_{3,1} & P_{3,2} & 1 & P_{3,4} \\ P_{4,1} & P_{4,2} & P_{4,3} & 1 \end{bmatrix} \qquad VARMAT = \begin{bmatrix} A_0 & A_1 & A_2 & A_3 \\ A_3 & A_0 & A_1 & A_2 \\ A_2 & A_3 & A_0 & A_1 \\ A_1 & A_2 & A_3 & A_0 \end{bmatrix} \qquad (9.1)$$

$$ANDY = \begin{bmatrix} A_0 & A_1 & A_2 & A_3 \\ P_{2,1}A_3 & A_0 & P_{2,3}A_1 & P_{2,4}A_2 \\ P_{3,1}A_2 & P_{3,2}A_3 & A_0 & P_{3,4}A_1 \\ P_{4,1}A_1 & P_{4,2}A_2 & P_{4,3}A_3 & A_0 \end{bmatrix} \qquad SYD = \begin{bmatrix} S_0 & S_1 & S_2 & S_3 \\ P_{2,1}S_3 & S_0 & P_{2,3}S_1 & P_{2,4}S_2 \\ P_{3,1}S_2 & P_{3,2}S_3 & S_0 & P_{3,4}S_1 \\ P_{4,1}S_1 & P_{4,2}S_2 & P_{4,3}S_3 & S_0 \end{bmatrix} \qquad (9.2)$$

The third part of the program:

The next step is to calculate the algebraic matrix form by eliminating parameters until we have a multiplicatively closed form. We begin by forming a product matrix from ANDY and SYD.

> $PROD := evalm(ANDY\&*SYD)$:

We insist upon the product matrix being of the same form as the two matrices ANDY and SYD. For example, the $a_{4,1}$ element of the product matrix must be equal to $P_{4,1}a_{1,2}$. Although the matrices are generally non-commutative, the order of multiplication does not matter because we seek only the form of the matrix.

In the case of the cyclic groups, much, but not all, of this elimination can be done by computer code. That the elimination can be done by code regardless of order indicates an underlying repetitive structure within the cyclic groups. Unfortunately, this repetitive structure, although observed by your author, is not clearly understood by your author. We would expect a repetitive structure for every type of group because we are doing group theory, and it is probably your author's lack of understanding that leads to his being unable to see the repetitive structure in most types of groups. Undoubtedly, this area of mathematics calls out for the attention of specialist group theorists.

In most other groups, the elimination of parameters is laboriously done by hand by your author. Although the elimination of parameters under the leading diagonal is easily done by insisting upon the equality of the elements of the leading diagonal of the product matrix, this will take us only half of the way and the elimination of the parameters above the leading diagonal is often fraught with problems for higher order groups. Other than the cyclic groups, it is not known how to be sure that we proceed in the proper direction. Often, one attempt to eliminate parameters and come to a multiplicatively closed form will lead to a quadratic elimination equation whereas a different (better) approach will lead to a linear elimination equation. We cannot even be sure of the number of parameters we need to eliminate to get multiplicative closure or even if we can get multiplicative closure. Whether the origin of these problems is your author's inadequacy or some mathematical property of the group is unknown. Your author once held the opinion that such failures were due to human error, but he is now wondering if there is some mathematical reason for these failures.

Such elimination by hand is coded like:

> $P[3, 2] := \dfrac{P[4, 1]}{P[2, 3]}$: $P[4, 2] := \dfrac{P[3, 1]}{P[2, 4]}$: $P[5, 2] := \dfrac{P[6, 1]}{P[2, 5]}$:

> ...

> $P[5, 6] := \dfrac{P[2, 5] \cdot P[5, 1]}{P[6, 1]}$: $P[4, 6] := \dfrac{P[2, 5] \cdot P[4, 5] \cdot P[5, 1]^2}{P[2, 8] \cdot P[6, 1] \cdot P[8, 1]}$:

> ...

> $P[3, 1] := \dfrac{P[2, 5]^4 \cdot P[5, 1]^2}{P[2, 3]^2 \cdot P[6, 1]^2}$:

If, after such parameter elimination code, we have come to a multiplicatively closed algebraic matrix form, then we have now completed the work on the individual group.

Henceforward, all the code in the program will act upon the multiplicatively closed algebraic matrix form of the given group. The same computer code is used to analyse all groups.

The fourth part of the program:

Having obtained a 'multiplicity closed' algebraic matrix form we need to test that algebraic matrix form to be sure it is multiplicatively closed. In practice, this testing piece of code is useful in the elimination by hand task because it throws out the points of multiplicative non-closure.

We reiterate; this code, and the subsequent code, is common to all groups.

The code which tests for multiplicative closure is:

Test for multiplicative closure

#CLOSURE compares each element of the levelled product matrix with the top row of the product matrix. For multiplicative closure closure is zero

```
>  PROD := Matrix(SIZE) :  LEVELPROD := Matrix(SIZE) :
>  PROD := evalm(ANDY&*SYD) :
>
>  for xxx  from 1 to  SIZE do
>  for yyy  from 1 to  SIZE do
```

$$LEVELPROD[xxx, yyy] := \left(\frac{PROD[xxx, yyy]}{PARAM[xxx, yyy]} \right)$$

 # LEVELPROD is elememts of the product matrix divied by the local parameters.

```
>  end do; end do ;
>
>  for zzz from 1 to (SIZE) do
>  for xxx  from 1 to  SIZE do
>  for yyy  from 1 to  SIZE do
>  if VARMAT[xxx, yyy] = A[zzz − 1] then
>  CLOSURE := factor(simplify(LEVELPROD[xxx, yyy] − LEVELPROD[1, zzz]))
>  if CLOSURE ≠ 0 then print("WE DO NOT HAVE MULTIPLICATIVE CLOSURE"); end if;
>
>  #print("CLOSURE is ", CLOSURE);
```

 # Removing the hash will reveal the points of non-closure

```
>  end if;
>  end do; end do; end do;
```

End of test for Multiplicative Closure

Having obtained a confirmed algebraic matrix form for the particular group we formed in the first part of the program with the SIZE variable and throughout the second and third parts of the program, we prepare to begin to analyse the algebras within that algebraic matrix form. This preparation is gathering together information about the group, gathering the remaining parameters into a list, and producing a list of all possible permutations of the parameters as ±1. We first need discover and store the orders of the elements of the group and the list of parameters which we have not eliminated. We do this in the next two (the fifth and sixth) parts of the program.

The fifth part of the program:

We first calculate the orders of each element of the group. The order of a group element is how many times we have to multiply it by itself to get the identity element of the group. The order of a group element corresponds to the power of the root of plus or minus unity that will correspond to that group element in the division algebras of that group. For example, if an element of the group, x, is of order three such that $x^3 = identity$, the that element will correspond to a cubic root of plus unity or to a cubic root of minus unity in all the division algebras of that group. The code that calculates the orders of each element of the group is:

We will now calculate the root distribution

```
> RLIST := [ ] :
      # We define a list to hold the roots. This list wil also be used later to form the individual
      algebras

> for subscrip from 0 to (SIZE − 1) do
> X[subscrip] := 0                    # We set the variables in VARMAT2 to zero
> end do:
> for subscrip from 1 to (SIZE − 1) do
> X[subscrip] := 1 :
> for pow from 2 to SIZE do
> RTEST := evalm(VARMAT2^pow)
> if RTEST[1, 1] = 1 then
>   RLIST := [op(RLIST), A[subscrip]]; RLIST := [op(RLIST), pow]
>   X[subscrip] := 0;
> end if: end do: end do:                # We now have the list of root powers
> for rootnum from 1 to (2· SIZE − 2) by 2 do
> print("The variable ", RLIST[rootnum], " is a ", RLIST[rootnum + 1], " root" )
> end do;
> print("RLIST is ", RLIST)
```

We now have the powers of the roots in a list called RLIST

We will need the information stored in RLIST later in the program. The above code prints out, using the cyclic group C_6 as an example, as:

The variable A_1 is a 6th root.

The variable A_2 is a 3th root.

The variable A_3 is a 2th root.

The variable A_4 is a 3th root.

The variable A_5 is a 6th root.

RLIST is $[A_1, 6, A_2, 3, A_3, 2, A_4, 3, A_5, 6]$

The sixth part of the program:

We store the parameters we did not earlier eliminate into a list called PARAMLIST. The code that does this is:

Stop press: The fourth line of this part of the program reads
if $length(PARAM[row, col]) = 9$ **then**
it should read
if $length(PARAM[row, col]) = 9$ **or** $length(PARAM[row, col]) = 10$ **then**

We will collect into a list all the parameters

> $PARAMLIST := [\ ]$:
> **for** *row* **from** 1 **to** *SIZE* **do**
>   *# We pick out the unelimated parameters based on their length, which is 9*

> **for** *col* **from** 1 **to** *SIZE* **do**
> **if** $length(PARAM[row, col]) = 9$ **then**
> $PARAMLIST := [op(PARAMLIST), P[row, col]]$
> **end if**: **end do**: **end do**:
> $PARAMLIST := MakeUnique(PARAMLIST)$:
> $print("PARAMLIST\ is\ ", PARAMLIST)$;

We now have a list of parameters called PARAMLIST

In the case of the order six cyclic group, this prints out as:

$$\text{PARAMLIST is } \left[P_{2,1}, P_{2,3}, P_{2,4}, P_{3,1}, P_{4,1} \right]$$

We are now prepared to start our analysis of the division algebras of the group.

The seventh part of the program:

In this part of the program, we construct a list of all possible permutations of the parameters as ± 1. The code that does this is:

We now form a list of possible Permutations (of the parameters).

> $PERM := [\ ]$:
> $PARAMSIGNS := [\ ]$:
>   *# This will hold every permutation of plus unity and minus unity of length (SIZE-1).*

>
> **for** *xx* **from** 1 **to** $numelems(PARAMLIST)$ **do**
>   *# This next piece of code puts (SIZE-1) minus ones and (SIZE-1) plus ones into PERM*

> $PERM := [op(PERM), +1]$
> **end do**:
> **for** *xx* **from** 1 **to** $numelems(PARAMLIST)$ **do**
> $PERM := [op(PERM), -1]$

> **end do**:
>> # *We now have enough +1s and -1s in PERM to form all permutations of +1 & -1. Each*
>> *permutation corresponds to a particular algebra.*

>

> *PARAMSIGNS* := *permute*(*PERM*, *numelems*(*PARAMLIST*)) :
>> # *This forms all possible permutations of (SIZE-1) +1s and -1s*

> #*print*("PARAMSIGNS is ", *PARAMSIGNS*) :

If we let this print PARAMSIGNS, we get a list of the permutations of n parameters equal to plus or minus unity. This list looks like:

$$\text{PARAMSIGNS is } \left[\,[1,1,1,1,1],[1,1,1,1,-1],[1,1,1,-1,1],\ldots\ \ldots[-1,-1,-1,-1,-1]\,\right]$$

We will call upon this list to produce each separate algebra.

The eighth part of the program:

We are now ready to examine each algebra and to test its determinant as we seek for the emergent distance functions of geometric spaces. We begin by defining the objects we will need.

We now start the Analysis

> *ROOTLIST* := [ ] :
>> # *ROOTLIST will store the plus or minus unity roots of an algebra in a list of (SIZE-1)*

> *LIST_OF_ROOTLISTS* := [ ] :
>> # *LIST_OF_ROOTLISTS is a list of the roots of each algebra*

> *COMM* := *Matrix*(*SIZE*) :   # *This matrix will test for commutativity*
> *FRED* := *Matrix*(*SIZE*) :   # *FRED is a temporary matrix*
> *PARAMCOPY* := *Matrix*(*SIZE*) :   # *PARACOPY is a temporary matrix*
> *THISALG* := *Matrix*(*SIZE*) :   # *This matrix will be each particular algebra*
> *LIST_OF_DETERMINANTS* := [ ] :   # *This list will hold all determinants.*
> *COMM_LIST* := [ ] :
>> # *COMM_LIST will store the commutativity of each algebra*

We now gather the data we will require. We now form a list of every algebra, whether it is commutative, and its determinant. This is done in one large loop (the *yy* loop) which visits each algebra in turn. The code that does this is:

> **for** *yy* **from** 1 **to** *numelems*(*PARAMSIGNS*) **do**
>> # *This loop extracts the imaginary roots (+1 or -1) of every algebra and puts them into*
>> *LIST_OF_ROOTLISTS*

> *PARAMCOPY* := *copy*(*PARAM*) :
>

We form the parameter matrix of each algebra.

>   **for** *zz* **from** 1 **to** *numelems*(*PARAMLIST*) **do**
>   *#These three loops form the PARAM matrix for each algebra*

>   **for** *row* **from** 1 **to** *SIZE* **do**
>   **for** *col* **from** 1 **to** *SIZE* **do**
>   *PARAMCOPY*[*row*, *col*] := *subs*({*PARAMLIST*[*zz*] = *PARAMSIGNS*[*yy*, *zz*]},
>   *PARAMCOPY*[*row*, *col*])

>   **end do**; **end do**; **end do**;
>   *# zz loop   # At this point we have a matrix of +1s & -1s corresponding to the parameter.*
>   *of a particular algebra*

>   *#print*("PARAMCOPY is ", *PARAMCOPY*);
>

We use the individual parameter matrix to form the individual algebraic matrix form.

>   **for** *row* **from** 1 **to** *SIZE* **do**
>   *# This bit of code constructs each of the separate algebras one at a time*

>   **for** *col* **from** 1 **to** *SIZE* **do**
>   *THISALG*[*row*, *col*] := *VARMAT3*[*row*, *col*]·*PARAMCOPY*[*row*, *col*] :
>   **end do**; **end do**;
>   *# We now have this particular algebra in a matrix called THISALG*

>

We test if the algebra is commutative or not and save the answers.

>   *COMMFLAG* := "COMMUTATIVE" :
>   *# This code tests for commutativity. We will attach this flag to each algebra in*
>   *LIST_OF_ROOTLISTS*

>   *COMM* := *evalm*(*ANDY*&*SYD − *SYD*&*ANDY) :
>   **for** *row* **from** 1 **to** *SIZE* **do**
>   **for** *col* **from** 1 **to** *SIZE* **do**
>   **if** *COMM*[*row*, *col*] ≠ 0 **then** *COMMFLAG* := "NON-COMMUTATIVE" : **end if**:
>   **end do**: **end do**:

>   *COMM_LIST* := [*op*(*COMM_LIST*), *COMMFLAG*] :
>   *# Because lists are ordered, we can match each algebra in LIST_OF_ROOTLISTS with*
>   *each element of COMM_LIST.*

>   *#print*("COMMFLAG is ", *COMMFLAG*) :
>

We prepare to find whether the roots of the individual algebra are roots of plus unity or of minus unity.

>   **for** *zz* **from** 0 **to** (*SIZE* − 1) **do**
>   *# We set the variables in VARMAT2 to zero. We are preparing to find whether the roots are*
>   *plus unity or minus unity.*

>   *X*[*zz*] := 0
>   **end do**:
>

>   **for** *subscrip* **from** 1 **to** (*SIZE* − 1) **do**
>   *# We set one variable in VARMAT2 to one. We will examine each root.*

>   *X*[*subscrip*] := 1 :

>

We find the signs of the roots.

> **for** *row* **from** 1 **to** *SIZE* **do**   # *This forms a matrix with only one non-zero variable.*
> **for** *col* **from** 1 **to** *SIZE* **do**
> $FRED[row, col] := VARMAT2[row, col] \cdot PARAMCOPY[row, col]$
> **end do; end do;**
>
> $pow := RLIST[2 \cdot subscrip];$
> $FREDPOW := evalm(FRED^{pow}) :$
> **if** $FREDPOW[1, 1] = 1$ **then** $ROOTLIST := [op(ROOTLIST), \text{alpha}]$; **end if:**
>   # *We use the symbol alpha to represent plus unity*
>
> **if** $FREDPOW[1, 1] = -1$ **then** $ROOTLIST := [op(ROOTLIST), \text{beta}]$; **end if:**
>   # *We use the symbol beta to represent minus unity*
>
> $X[subscrip] := 0 :$
> **end do:**
>   # *subscript loop*   # *We now have the set of roots in*
>   *ROOTLIST*
>
> $LIST\_OF\_ROOTLISTS := [op(LIST\_OF\_ROOTLISTS), ROOTLIST] :$
>   # *We save this set of roots in a list.*

We now know the signs of the roots.

>

We construct each algebra to be attached to each determinant.

> $ALGEB := A[0] :$
> **for** *PROOT* **from** 1 **to** $(SIZE - 1)$ **do**
>   # *This loop identifies each algebra to be added into the determinant lists below*
>
> $ALGEB := ALGEB + A[PROOT] \cdot (ROOTLIST[PROOT])^{\left(\frac{1}{RLIST[2 \cdot PROOT]} \right)}$
> **end do:**
> #*print*("ALGEB is ", *ALGEB*);
>

We take the determinant (distance function) of the algebra and save it.

> $THISALGDET := Determinant(THISALG) :$ # *We collect the determinants into a list*
> $LIST\_OF\_DETERMINANTS := [op(LIST\_OF\_DETERMINANTS), THISALGDET, ALGEB] :$
>
> $ROOTLIST := [\ ] :$ #*We clean out the rootlist*
> **end do:**
>   # *yy loop.   We now have every algebra's roots in LIST_OF_ROOTLISTS. We have every*
>   *algebras determinant*

We now have the data we require. We can immediately print out some of the analysis.

print("There are ", *numelems*(*LIST_OF_ROOTLISTS*) , "separate algebras")

This prints out as, in the case of the group C_6 :

52

There are 32 separate algebras.

We have now counted the number of algebras in the given group. We could have done it by simply raising 2 to the power of the number of parameters. Note that many of these separate algebras will be isomorphic to each other.

We can also print out the list of root lists:

> *print*("alpha represents plus unity and beta represents minus unity");
> *print*("The list of root-lists is: ", *LIST_OF_ROOTLISTS*);

This prints out as, in the case of the group C_6:

alpha represents plus unity and beta represents minus unity

The list of root-lists is: $\left[[\alpha,\alpha,\alpha,\alpha,\alpha], [\beta,\beta,\beta,\beta,\beta], [\alpha,\alpha,\alpha,\beta,\alpha], ... \right]$

We can print out the commutativity list.

print("COMM_LIST is ", *COMM_LIST*) :

In the case of the group C_6 this is:

COMM_LIST is: $[COMMUTATIVE, COMMUTATIVE, ...]$

If we wish, we can print out the list of determinants:

> *print*("There are ",
>   numelems(*LIST_OF_DETERMINANTS*) " elements in the
>   LIST_OF_DETERMINANTS") :
> *print*("The list of determinants is ", *LIST_OF_DETERMINANTS*) :

In the case of the group C_6 this is:

There are 64 elements in the LIST_OF_DETERMINANTS

The list of determinants is: $\left[B_0^6 - 6B_0^4 B_1 B_5 - ... - B_5^6, A_0 + A_1\alpha^{\frac{1}{6}} + A_2\alpha^{\frac{1}{3}} + A_3\sqrt{\alpha} + A_4\alpha^{\frac{1}{3}} + A_5\alpha^{\frac{1}{6}}, B_0^6 - ..... \right]$

There are 64 elements in the determinant list of the group C_6 because each of the 32 determinants is attached to an algebra.

The ninth part of the program:
We now gather the different algebras together and present them. The code that does this is:

We form a list of the separate algebras.
> *ALGLIST* := [ ] :   # *ALGLIST will hold the separate algebras.*

```
> for zz from 1 to numelems(LIST_OF_ROOTLISTS) do
> ALG := A[0];
> for yy from 1 to (SIZE − 1) do
```

$$> \quad ALG := ALG + A[yy] \cdot (LIST\_OF\_ROOTLISTS[zz, yy])^{\left(\frac{1}{RLIST[2 \cdot yy]}\right)}$$

```
> end do:
> ALG := cat(ALG, COMM_LIST[zz]) :   # We add in the commutative status of the algebra
> ALGLIST := [op(ALGLIST), ALG]
> end do:
> #print(ALGLIST);
>
```

We prepare to count the numbers of isomorphic algebras.

```
> for zz from 1 to (SIZE − 1) do
        # We put the variables in the algebras equal because we are interested in the types of roots
        only
> A[zz] := A[1]
> end do:
>
```

We put all the algebras and the numbers of occurrences of them into a summary list and cancel the duplicates.

```
> ALGLISTSUMMARY := [ ] :
        # ALGLISTSUMMARY will hold all the algebras and the numbers of them

> for zz from 1 to numelems(ALGLIST) do
> ALGLISTSUMMARY := [op(ALGLISTSUMMARY), cat(Occurrences(ALGLIST[zz],
        ALGLIST ) , ALGLIST[zz]) ]

> #print(Occurrences(ALGLIST[zz] , ALGLIST ) , ALGLIST[zz])
> end do:
> ALGLISTSUMMARY := MakeUnique(ALGLISTSUMMARY) :
>
> print("The group ", THE_GROUP, "has the following non-isomorphic algebras within it." )
> print("There are ", numelems(ALGLISTSUMMARY) , " non-isomorphic algebras")
> for zz from 1 to numelems(ALGLISTSUMMARY) do
> print(ALGLISTSUMMARY[zz])
> end do;
```

This prints out, in the case of the group C_6 as:

The group Cyclic order 6 has the following non-isomorphic algebras within it.

There are 6 non-isomorphic algebras.

$$4\| \| \left(A_0 + 2A_1\alpha^{\frac{1}{6}} + 2A_1\alpha^{\frac{1}{3}} + A_1\sqrt{\alpha}\right) \| \text{``COMMUTATIVE''}$$

$$4\| \| \left(A_0 + 2A_1\beta^{\frac{1}{6}} + 2A_1\beta^{\frac{1}{3}} + A_1\sqrt{\beta}\right) \| \text{``COMMUTATIVE''}$$

$$8\| \ \| \ \left(A_0 + 2A_1\alpha^{\frac{1}{6}} + A_1\alpha^{\frac{1}{3}} + A_1\sqrt{\alpha} + A_1\beta^{\frac{1}{3}} \right) \ \| \ \text{``COMMUTATIVE''}$$

$$8\| \ \| \ \left(A_0 + 2A_1\beta^{\frac{1}{6}} + A_1\beta^{\frac{1}{3}} + A_1\sqrt{\beta} + A_1\alpha^{\frac{1}{3}} \right) \ \| \ \text{``COMMUTATIVE''}$$

$$4\| \ \| \ \left(A_0 + 2A_1\alpha^{\frac{1}{6}} + 2A_1\beta^{\frac{1}{3}} + A_1\sqrt{\alpha} \right) \ \| \ \text{``COMMUTATIVE''}$$

$$4\| \ \| \ \left(A_0 + 2A_1\beta^{\frac{1}{6}} + 2A_1\alpha^{\frac{1}{3}} + A_1\sqrt{\beta} \right) \ \| \ \text{``COMMUTATIVE''}$$

We have now analysed the algebras within the given group.

The tenth and final part of the program:

We now look through the determinants of these algebras seeking a geometric space. If the determinants do not factorise to quadratic form, then there will be no geometric space supporting 2-dimensional rotations. Ditto factorisation to cubic form and geometric spaces supporting 3-dimensional rotations. We know what the factorisation of the determinant will have to be like to produce a geometric space supporting 2-dimensional rotations; it will have to be of the form:

$$\det{}^n = \left(a^2 \pm b^2 \pm c^2 \pm \ldots \right)^{\frac{n}{2}} \qquad (9.3)$$

We can print out the factorised determinants and quickly inspect the determinants by eye. The code that does this is:

We print out the factorised determinants.

```
>  for zz from 1 to  numelems( LIST_OF_DETERMINANTS) by 2 do;
>  FACDET := factor(LIST_OF_DETERMINANTS[zz]) :
>  print( FACDET)
>  end do:
```

Summary:

In part one of the program we have to enter the SIZE variable as the order of the group and specify THE_GROUP variable.

In parts two and three, we have to form the algebraic matrix form of the group ready to be fed into the subsequent parts of the program.

When we have the algebraic matrix form, the rest of the program will analyse the group algebras and print out the factorised determinants. We inspect the factorised determinants by eye looking for geometric spaces.

SECTION II – The Cyclic Groups

Chapter 10

The Cyclic Groups in General

All cyclic groups have a Standard form Cayley table which has variables positioned in parallel to the leading diagonal identity and 'marching across the page'. For examples, see the Cayley tables of the various cyclic groups within this book.

In the case of the cyclic groups, the calculations we do to construct the algebraic matrix forms are done by computer. Your author uses the Maple 17 programming language.

The computer program for the cyclic groups:
We set the parameters and the Cayley table variables at the same time. We produce two matrices called ANDY and SYD. We also produce several copies of the matrices VARMAT and PARAM which we will need later. VARMAT is the variables without parameters (the Cayley table). PARAM is the parameters without the variables. Although we use Latin letters for variables in this book, the computer uses A_n for variables where $n = 0, 1, 2, \ldots$. In Maple 17, the code that does this is:

This program deals with the cyclic groups
We set the order of the cyclic group with the SIZE variable

```
>  restart :
>  with(LinearAlgebra) : with(ListTools) : with(combinat) :
>  SIZE := 6;
        # SIZE is the order of the cyclic group

>  THE_GROUP := "Cyclic order 6" :
        # We make a note of which group we are working with for presentation later
```

We now set up the basic matrices of the cyclic group.

```
>  dance := 0 :
        # We install the variables  and the parameters of the cyclic group
>  for row from 1 to SIZE do
>  dance := dance + 1 :
>  for col from 1 to SIZE do
>  VARMAT[row, col] := A[(col − dance) mod SIZE]
>  VARMAT2[row, col] := X[(col − dance) mod SIZE]
>  VARMAT3[row, col] := B[(col − dance) mod SIZE]
```

> $VARMAT4[row, col] := C[(col - dance) \bmod SIZE]$
> $ANDY[row, col] := P[row, col] \cdot A[(col - dance) \bmod SIZE]$
> $SYD[row, col] := P[row, col] \cdot S[(col - dance) \bmod SIZE]$
> $PARAM[row, col] := P[row, col]$:
> **end do**;
> **end do**:   # Matrices are now set up
> **for** *col* **from** 1 **to** *SIZE* **do**
>   # We put the top row parameters to unity and we put the leading diagonal parameters to
>   unity
> $P[1, col] := 1$;
> $P[col, col] := 1$;
> **end do**:   # We are now ready to begin eliminating parameters
>
> #print(ANDY, PARAM, VARMAT) :
>   # If we want to see the result of our work, we print that result.

We illustrate the procedure using the order five and order six cyclic groups. We now have:

$$
C_5 = \begin{bmatrix}
a & b & c & d & e \\
P_{2,1}e & a & P_{2,3}b & P_{2,4}c & P_{2,5}d \\
P_{3,1}d & P_{3,2}e & a & P_{3,4}b & P_{3,5}c \\
P_{4,1}c & P_{4,2}d & P_{4,3}e & a & P_{4,5}b \\
P_{5,1}b & P_{5,2}c & P_{5,3}d & P_{5,4}e & a
\end{bmatrix}
\quad
C_6 = \begin{bmatrix}
a & b & c & d & e & f \\
P_{2,1}f & a & P_{2,3}b & P_{2,4}c & P_{2,5}d & P_{2,6}e \\
P_{3,1}e & P_{3,2}f & a & P_{3,4}b & P_{3,5}c & P_{3,6}d \\
P_{4,1}d & P_{4,2}e & P_{4,3}f & a & P_{4,5}b & P_{4,6}c \\
P_{5,1}c & P_{5,2}d & P_{5,3}e & P_{5,4}f & a & P_{5,6}b \\
P_{6,1}b & P_{6,2}c & P_{6,3}d & P_{6,4}e & P_{6,5}f & a
\end{bmatrix}
$$

(10.1)

We begin the calculation of the algebraic matrix form by insisting upon the equality of the elements on the leading diagonal of the product matrix. This is just insisting upon multiplicative closure of matrix form. In Maple 17, the code that does this is:s

> **for** *col* **from** 1 **to** *SIZE* **do**
>   # This calculate the parameters under the leading diagonal except fot the leftmost column

> **for** *row* **from** 3 **to** *col* **do**
> $P[col, (row - 1)] := \dfrac{P[col + 2 - row, 1]}{P[row - 1, col]}$:

> **end do**; **end do**;

This gives:

$$C_5 = \begin{bmatrix} a & b & c & d & e \\ P_{2,1}e & a & P_{2,3}b & P_{2,4}c & P_{2,5}d \\ P_{3,1}d & \dfrac{P_{2,1}}{P_{2,3}}e & a & P_{3,4}b & P_{3,5}c \\ P_{4,1}c & \dfrac{P_{3,1}}{P_{2,4}}d & \dfrac{P_{2,1}}{P_{3,4}}e & a & P_{4,5}b \\ P_{5,1}b & \dfrac{P_{4,1}}{P_{2,5}}c & \dfrac{P_{3,1}}{P_{3,5}}d & \dfrac{P_{2,1}}{P_{4,5}}e & a \end{bmatrix}$$

$$C_6 = \begin{bmatrix} a & b & c & d & e & f \\ P_{2,1}f & a & P_{2,3}b & P_{2,4}c & P_{2,5}d & P_{2,6}e \\ P_{3,1}e & \dfrac{P_{2,1}}{P_{2,3}}f & a & P_{3,4}b & P_{3,5}c & P_{3,6}d \\ P_{4,1}d & \dfrac{P_{3,1}}{P_{2,4}}e & \dfrac{P_{2,1}}{P_{3,4}}f & a & P_{4,5}b & P_{4,6}c \\ P_{5,1}c & \dfrac{P_{4,1}}{P_{2,5}}d & \dfrac{P_{3,1}}{P_{3,5}}e & \dfrac{P_{2,1}}{P_{4,5}}f & a & P_{5,6}b \\ P_{6,1}b & \dfrac{P_{5,1}}{P_{2,6}}c & \dfrac{P_{4,1}}{P_{3,6}}d & \dfrac{P_{3,1}}{P_{4,6}}e & \dfrac{P_{2,1}}{P_{5,6}}f & a \end{bmatrix}$$

(10.2)

Inspection of (10.2) reveals that the parameters which have been eliminated are every parameter under the leading diagonal except the leftmost column. Looking at the code above, we see that this calculation has been done, not by solving parameter elimination equations but by installing an observed pattern. Clearly, there is a reason for this pattern. We look at the a_{22} element and the a_{33} element of the product of two C_5 matrices; we consider only the eb variables in these elements. We have:

$$a_{22} = ... + \left(P_{2,1} + P_{2,3}P_{3,2} \right) eb + ...$$
$$a_{33} = ... + \left(P_{3,4}P_{4,3} + P_{2,3}P_{3,2} \right) eb + ...$$

(10.3)

Since these have to be equal, we have $P_{2,1} = P_{3,4}P_{4,3}$.

Examination of the algebraic matrix forms of the cyclic groups presented in this book shows a reflective symmetry down each of the columns above the leading diagonal. For example; the C_5 algebraic matrix form, (11.2), has columns above the leading diagonal:

$$[1] \begin{bmatrix} 1 \\ 1 \end{bmatrix} \begin{bmatrix} 1 \\ P_{2,3} \\ 1 \end{bmatrix} \begin{bmatrix} 1 \\ P_{2,4} \\ P_{2,4} \\ 1 \end{bmatrix} \begin{bmatrix} 1 \\ \dfrac{P_{2,3}P_{3,1}}{P_{2,1}} \\ \dfrac{P_{2,4}P_{3,1}}{P_{2,1}} \\ \dfrac{P_{2,3}P_{3,1}}{P_{2,1}} \\ 1 \end{bmatrix}$$

(10.4)

The next step is to impose this reflective symmetry on the algebraic matrix form above the leading diagonal. The code that does this is:

```
> for col from 4 to SIZE do
       # This installs reflective symmetry above the leading diagonal
```

> **for** *row* **from** $(col - 1)$ **to** floor$\left(\dfrac{(col - 1)}{2} \right)$ **by** (-1) **do**

> $P[row, col] := P[col - row + 1, col]$:

> **end do**; **end do**;

We now have:

$$C_5 = \begin{bmatrix} a & b & c & d & e \\ P_{2,1}e & a & P_{2,3}b & P_{2,4}c & P_{2,5}d \\ P_{3,1}d & \dfrac{P_{2,1}}{P_{2,3}}e & a & P_{2,4}b & P_{3,5}c \\ P_{4,1}c & d\dfrac{P_{3,1}}{P_{2,4}} & e\dfrac{P_{2,1}}{P_{3,4}} & a & P_{2,5}b \\ P_{5,1}b & \dfrac{P_{3,1}}{P_{2,5}}c & \dfrac{P_{3,1}}{P_{3,5}}d & \dfrac{P_{2,1}}{P_{4,5}}e & a \end{bmatrix}$$

$$C_6 = \begin{bmatrix} a & b & c & d & e & f \\ P_{2,1}f & a & P_{2,3}b & P_{2,4}c & P_{2,5}d & P_{2,6}e \\ P_{3,1}e & \dfrac{P_{2,1}}{P_{2,3}}f & a & P_{2,4}b & P_{3,5}c & P_{3,6}d \\ P_{4,1}d & \dfrac{P_{3,1}}{P_{2,4}}e & \dfrac{P_{2,1}}{P_{3,4}}f & a & P_{2,5}b & P_{3,6}c \\ P_{5,1}c & \dfrac{P_{4,1}}{P_{2,5}}d & \dfrac{P_{3,1}}{P_{3,5}}e & \dfrac{P_{2,1}}{P_{4,5}}f & a & P_{2,6}b \\ P_{6,1}b & \dfrac{P_{3,1}}{P_{2,6}}c & \dfrac{P_{4,1}}{P_{3,6}}d & \dfrac{P_{3,1}}{P_{4,6}}e & \dfrac{P_{2,1}}{P_{5,6}}f & a \end{bmatrix}$$

(10.5)

The parameters in the third row above the leading diagonal are given by:

$$P_{3,x} = \frac{PARAM_{2,x} PARAM_{2,(x-1)}}{P_{2,3}}$$

(10.6)

There is a hiccup with larger matrices which means we always have to start at column five. We install these parameters. The code that does this is:

> **if** *SIZE* < 7 **then** *STARTCOL* := ceil$\left(\dfrac{SIZE}{2} + 1 \right)$ **end if**:   # *This installs the third row*

> **if** *SIZE* ≥ 7 **then** *STARTCOL* := 5 : **end if**:

> **for** *col* **from** *STARTCOL* **to** $(SIZE - 1)$ **do**

> $P[3, col] := \dfrac{PARAM[2, col] \cdot PARAM[2, (col - 1)]}{P[2, 3]}$:

> **end do**:

We now have:

$$C_5 = \begin{bmatrix} a & b & c & d & e \\ P_{2,1}e & a & P_{2,3}b & P_{2,4}c & P_{2,5}d \\ P_{3,1}d & \dfrac{P_{2,1}}{P_{2,3}}e & a & P_{2,4}b & P_{3,5}c \\ P_{4,1}c & d\dfrac{P_{3,1}}{P_{2,4}} & e\dfrac{P_{2,1}}{P_{3,4}} & a & P_{4,5}b \\ P_{5,1}b & \dfrac{P_{3,1}}{P_{2,5}}c & \dfrac{P_{3,1}}{P_{3,5}}d & \dfrac{P_{2,1}}{P_{4,5}}e & a \end{bmatrix}$$

$$C_6 = \begin{bmatrix} a & b & c & d & e & f \\ P_{2,1}f & a & P_{2,3}b & P_{2,4}c & P_{2,5}d & P_{2,6}e \\ P_{3,1}e & \dfrac{P_{2,1}}{P_{2,3}}f & a & P_{3,4}b & \dfrac{P_{2,4}P_{2,5}}{P_{2,3}}c & P_{3,6}d \\ P_{4,1}d & \dfrac{P_{3,1}}{P_{2,4}}e & \dfrac{P_{2,1}}{P_{3,4}}f & a & P_{2,5}b & P_{4,6}c \\ P_{5,1}c & \dfrac{P_{4,1}}{P_{2,5}}d & \dfrac{P_{3,1}}{P_{3,5}}e & \dfrac{P_{2,1}}{P_{4,5}}f & a & P_{5,6}b \\ P_{6,1}b & \dfrac{P_{3,1}}{P_{2,6}}c & \dfrac{P_{4,1}}{P_{3,6}}d & \dfrac{P_{3,1}}{P_{4,6}}e & \dfrac{P_{2,1}}{P_{5,6}}f & a \end{bmatrix}$$

$$(10.7)$$

This, (10.7), is a poor example, but we have not the paper to shown a more effective example.

There is a modified reflective symmetry across each row above the leading diagonal. Looking at the algebraic matrix form of the C_9 group, (15.2), we see this is of the form:

$$P_{3,5} = \frac{P_{2,4}P_{2,5}}{P_{2,3}} \qquad \& \qquad P_{3,8} = \frac{P_{2,4}P_{2,5}}{P_{2,3}}\frac{P_{5,1}}{P_{3,1}} \qquad (10.8)$$

We install this symmetry. The code that does this is:

```
> for zz from 2 to SIZE/2 do   # zz Loop

>    for col from ceil( (SIZE + (zz + 2))/2 ) to (SIZE) do
        # This installs the Right side parameters based on a left to right modified symmetry

>    P[zz, col] := ( P[zz, (SIZE − col + zz + 1)]·P[(SIZE − col + zz + 1), 1] ) / ( P[(SIZE − col + 2), 1] ) :

>    end do:
> end do:   # zz Loop
```

We now have:

$$
C_5 = \begin{bmatrix}
a & b & c & d & e \\[2mm]
P_{2,1}e & a & P_{2,3}b & P_{2,4}c & \dfrac{P_{2,3}P_{3,1}}{P_{2,1}}d \\[4mm]
P_{3,1}d & \dfrac{P_{2,1}}{P_{2,3}}e & a & P_{2,4}b & P_{3,5}c \\[4mm]
P_{4,1}c & d\dfrac{P_{3,1}}{P_{2,4}} & e\dfrac{P_{2,1}}{P_{3,4}} & a & \dfrac{P_{2,3}P_{3,1}}{P_{2,1}}b \\[4mm]
P_{5,1}b & \dfrac{P_{3,1}}{P_{2,5}}c & \dfrac{P_{3,1}}{P_{3,5}}d & \dfrac{P_{2,1}}{P_{4,5}}e & a
\end{bmatrix}
$$

$$
C_6 = \begin{bmatrix}
a & b & c & d & e & f \\[2mm]
P_{2,1}f & a & P_{2,3}b & P_{2,4}c & \dfrac{P_{2,4}P_{4,1}}{P_{3,1}}d & \dfrac{P_{2,3}P_{3,1}}{P_{2,1}}e \\[4mm]
P_{3,1}e & \dfrac{P_{2,1}}{P_{2,3}}f & a & P_{2,4}b & \dfrac{P_{2,4}P_{2,4}P_{4,1}}{P_{2,3}P_{3,1}}c & \dfrac{P_{2,4}P_{4,1}}{P_{2,1}}d \\[4mm]
P_{4,1}d & \dfrac{P_{3,1}}{P_{2,4}}e & \dfrac{P_{2,1}}{P_{3,4}}f & a & \dfrac{P_{2,4}P_{4,1}}{P_{3,1}}b & \dfrac{P_{2,4}P_{4,1}}{P_{2,1}}c \\[4mm]
P_{5,1}c & \dfrac{P_{4,1}}{P_{2,5}}d & \dfrac{P_{3,1}}{P_{3,5}}e & \dfrac{P_{2,1}}{P_{4,5}}f & a & \dfrac{P_{2,3}P_{3,1}}{P_{2,1}}b \\[4mm]
P_{6,1}b & \dfrac{P_{3,1}}{P_{2,6}}c & \dfrac{P_{4,1}}{P_{3,6}}d & \dfrac{P_{3,1}}{P_{4,6}}e & \dfrac{P_{2,1}}{P_{5,6}}f & a
\end{bmatrix}
$$

(10.9)

For odd order groups, we are left with a parameter in the rightmost column, $P_{3,5}$ in the above C_5 matrix. We deal with this single parameter with the code:

> **if** *SIZE* **mod** $2 = 1$ **then**   # *Loop for* **odd** *order groups*

> $P\left[\left(\dfrac{SIZE+1}{2}\right), SIZE\right] := \dfrac{P\left[2, \text{ceil}\left(\dfrac{SIZE}{2}+1\right)\right] \cdot P\left[\left(\dfrac{SIZE+1}{2}\right), 1\right]}{P[2,1]}$:
> 　　# *This puts the middle row rightmost column parameter in*

> **end if**: # *loop for odd order groups*

We now install the reflective symmetry on the leftmost column. The code that does this is:

> **for** *row* **from** $\text{ceil}\left(\dfrac{SIZE}{2}\right) + 1$ **to** *SIZE* **do**
> 　　# *This calculates the parameters in the leftmost column*

> $P[row, 1] := P[(2 + SIZE - row), 1]$:
> **end do**:

We now have:

$$
C_5 = \begin{bmatrix}
a & b & c & d & e \\[4pt]
P_{2,1}e & a & P_{2,3}b & P_{2,4}c & \dfrac{P_{2,3}P_{3,1}}{P_{2,1}}d \\[10pt]
P_{3,1}d & \dfrac{P_{2,1}}{P_{2,3}}e & a & P_{2,4}b & P_{3,5}c \\[10pt]
P_{3,1}c & d\dfrac{P_{3,1}}{P_{2,4}} & e\dfrac{P_{2,1}}{P_{3,4}} & a & \dfrac{P_{2,3}P_{3,1}}{P_{2,1}}b \\[10pt]
P_{2,1}b & \dfrac{P_{3,1}}{P_{2,5}}c & \dfrac{P_{3,1}}{P_{3,5}}d & \dfrac{P_{2,1}}{P_{4,5}}e & a
\end{bmatrix}
$$

$$
C_6 = \begin{bmatrix}
a & b & c & d & e & f \\[4pt]
P_{2,1}f & a & P_{2,3}b & P_{2,4}c & \dfrac{P_{2,4}P_{4,1}}{P_{3,1}}d & \dfrac{P_{2,3}P_{3,1}}{P_{2,1}}e \\[10pt]
P_{3,1}e & \dfrac{P_{2,1}}{P_{2,3}}f & a & P_{2,4}b & \dfrac{P_{2,4}P_{2,4}P_{4,1}}{P_{2,3}P_{3,1}}c & \dfrac{P_{2,4}P_{4,1}}{P_{2,1}}d \\[10pt]
P_{4,1}d & \dfrac{P_{3,1}}{P_{2,4}}e & \dfrac{P_{2,1}}{P_{3,4}}f & a & \dfrac{P_{2,4}P_{4,1}}{P_{3,1}}b & \dfrac{P_{2,4}P_{4,1}}{P_{2,1}}c \\[10pt]
P_{3,1}c & \dfrac{P_{4,1}}{P_{2,5}}d & \dfrac{P_{3,1}}{P_{3,5}}e & \dfrac{P_{2,1}}{P_{4,5}}f & a & \dfrac{P_{2,3}P_{3,1}}{P_{2,1}}b \\[10pt]
P_{2,1}b & \dfrac{P_{3,1}}{P_{2,6}}c & \dfrac{P_{4,1}}{P_{3,6}}d & \dfrac{P_{3,1}}{P_{4,6}}e & \dfrac{P_{2,1}}{P_{5,6}}f & a
\end{bmatrix}
$$

(10.10)

However, this is not the end of it because for higher order groups, (14.2), there are still parameters above the leading diagonal to be eliminated. Until we have an understanding of a pattern, we have to eliminate the remaining parameters by hand. Those parameters are given in the code:

> **if** *SIZE* = 9 **then** $\quad P[4,7] := \dfrac{P[2,5]\cdot P[2,6]\cdot P[2,7]}{P[2,3]\cdot P[2,4]}$: **end if**:

> **if** *SIZE* = 10 **then** $\quad P[4,7] := \dfrac{P[2,5]\cdot P[2,6]\cdot P[2,7]}{P[2,3]\cdot P[2,4]}$: **end if**:

> **if** *SIZE* = 11 **then**
> $P[4,7] := \dfrac{P[2,5]\cdot P[2,6]\cdot P[2,7]}{P[2,3]\cdot P[2,4]}$: $P[4,8] := \dfrac{P[2,6]\cdot P[2,7]\cdot P[2,8]}{P[2,3]\cdot P[2,4]}$:
> **end if**:
>
> **if** *SIZE* = 12 **then**
> $P[4,7] := \dfrac{P[2,5]\cdot P[2,6]\cdot P[2,7]}{P[2,3]\cdot P[2,4]}$: $P[4,8] := \dfrac{P[2,6]\cdot P[2,7]\cdot P[2,8]}{P[2,3]\cdot P[2,4]}$: $P[5,9]$
> $\quad := \dfrac{P[2,6]^2\cdot P[2,7]^2\cdot P[7,1]}{P[2,3]\cdot P[2,4]\cdot P[2,5]\cdot P[5,1]}$:

> **end if**:
> **if** *SIZE* = 13 **then**
> $P[4,7] := \dfrac{P[2,5]\cdot P[2,6]\cdot P[2,7]}{P[2,3]\cdot P[2,4]}$: $P[4,8] := \dfrac{P[2,6]\cdot P[2,7]\cdot P[2,8]}{P[2,3]\cdot P[2,4]}$: $P[4,9]$
> $\quad := \dfrac{P[2,7]^2\cdot P[2,8]\cdot P[7,1]}{P[2,3]\cdot P[2,4]\cdot P[6,1]}$: $P[5,9] := \dfrac{P[2,6]\cdot P[2,7]^2 P[2,8]\cdot P[7,1]}{P[2,3]\cdot P[2,4]\cdot P[2,5]\cdot P[6,1]}$:

> **end if**:
>
> **if** *SIZE* = 14 **then**

>
$$P[4,7] := \frac{P[2,5] \cdot P[2,6] \cdot P[2,7]}{P[2,3] \cdot P[2,4]} \; : \quad P[4,8] := \frac{P[2,6] \cdot P[2,7] \cdot P[2,8]}{P[2,3] \cdot P[2,4]} \; : P[4,9]$$

$$:= \frac{P[2,7] \cdot P[2,8]^2 \cdot P[8,1]}{P[2,3] \cdot P[2,4] \cdot P[7,1]} \; : \quad P[5,9] := \frac{P[2,6] \cdot P[2,7] \, P[2,8]^2 \cdot P[8,1]}{P[2,3] \cdot P[2,4] \cdot P[2,5] \cdot P[7,1]} \; : P[5,10]$$

$$:= \frac{P[2,7]^2 \, P[2,8]^2 \cdot P[8,1]}{P[2,3] \cdot P[2,4] \cdot P[2,5] \cdot P[6,1]} \; :$$

> **end if**:
>
> **if** *SIZE* = 15 **then**
>
$$P[4,7] := \frac{P[2,5] \cdot P[2,6] \cdot P[2,7]}{P[2,3] \cdot P[2,4]} \; : \quad P[4,8] := \frac{P[2,6] \cdot P[2,7] \cdot P[2,8]}{P[2,3] \cdot P[2,4]} \; : P[4,9]$$

$$:= \frac{P[2,7] \cdot P[2,8] \cdot P[2,9]}{P[2,3] \cdot P[2,4]} \; : P[4,10] := \frac{P[2,8]^2 \, P[2,9] \cdot P[8,1]}{P[2,3] \cdot P[2,4] \cdot P[7,1]} \; : \quad P[5,9]$$

$$:= \frac{P[2,6] \cdot P[2,7] \, P[2,8] \cdot P[2,9]}{P[2,3] \cdot P[2,4] \cdot P[2,5]} \; :$$

>
$$P[5,10] := \frac{P[2,7] \, P[2,8]^2 \cdot P[2,9] \cdot P[8,1]}{P[2,3] \cdot P[2,4] \cdot P[2,5] \cdot P[7,1]} \; : \quad P[6,11]$$

$$:= \frac{P[2,7]^2 \, P[2,8]^2 \cdot P[2,9] \cdot P[8,1]}{P[2,3] \cdot P[2,4] \cdot P[2,5] \cdot P[2,6] \cdot P[6,1]} \; :$$

> **end if**:

We make no attempt at higher order cyclic groups. (My computer is not powerful enough.)

At this point in the program, we have created the algebraic matrix form of the cyclic groups. We append the testing and analysis code given as the fourth and subsequent parts of the computer program beginning with the test for multiplicative closure in the previous chapter. The computer program will now print out the algebras and determinants of the particular cyclic group.

Summary:
The above computer program presents the general way in which we like to do this mathematics. We cannot always install the variables and parameters with computer code, and we often have to do these things by hand. We will not present such detailed program for the other types of groups. The program enables us to examine the cyclic groups up to and including order fifteen. We are limited by the power of your author's PC.

The Order Five Cyclic Group

For any prime number, there is only one group of that order and that group is a cyclic group. Since the order of a subgroup must divide the order of the group, prime order groups have no subgroups. The only order five group is C_5. C_5 is a commutative group. The Standard form Cayley table of C_5 is:

$$C_5 = \begin{bmatrix} a & b & c & d & e \\ e & a & b & c & d \\ d & e & a & b & c \\ c & d & e & a & b \\ b & c & d & e & a \end{bmatrix} \tag{11.1}$$

There are other Standard form Cayley tables, but they produce only the same algebras up to swapping of variables. These other Standard form Cayley tables can be found as sums of the 5×5 permutation matrices with no entries on the leading diagonal (except for the identity) which are matrices of all 1's.

With parameters the C_5 algebraic matrix form is:

$$C_5 = \begin{bmatrix} a & b & c & d & e \\ P_{2,1}e & a & P_{2,3}b & P_{2,4}c & \dfrac{P_{2,3}P_{3,1}}{P_{2,1}}d \\ P_{3,1}d & \dfrac{P_{2,1}}{P_{2,3}}e & a & P_{2,4}b & \dfrac{P_{2,4}P_{3,1}}{P_{2,1}}c \\ P_{3,1}c & d\dfrac{P_{3,1}}{P_{2,4}} & e\dfrac{P_{2,1}}{P_{2,4}} & a & \dfrac{P_{2,3}P_{3,1}}{P_{2,1}}b \\ P_{2,1}b & \dfrac{P_{2,1}}{P_{2,3}}c & \dfrac{P_{2,1}}{P_{2,4}}d & \dfrac{P_{2,1}P_{2,1}}{P_{2,3}P_{3,1}}e & a \end{bmatrix} \tag{11.2}$$

This matrix, (11.2), is commutative, and all the algebras which it represents are commutative.

There are $2^4 = 16$ permutations of the four parameters as ± 1 giving sixteen division algebras. We remind the reader that we have to take the exponential of the above algebraic matrix form, (11.2), to get the division algebras.

Since five is a prime number, no distance functions will factorise to quadratic or cubic form, and so we have no emergent spaces which are geometric spaces.

There is one algebra with four fifth roots of plus unity:

$$1 \quad \text{of} \quad a + b\sqrt[5]{+1} + c\sqrt[5]{+1} + d\sqrt[5]{+1} + e\sqrt[5]{+1} \tag{11.3}$$

$$d^5 = a^5 + b^5 + c^5 + d^5 + e^5 - 5a^3bd + \ldots\ldots - 5cd^3e \tag{11.4}$$

We do not give the complete distance function, which will not factorise into a simpler expression, because it contains 26 terms. Such functions are best handled by a computer using the above algebraic matrix form, (11.2). We remind the reader that the distance function (norm) of a division algebra is the determinant of the algebraic matrix form.

There are four algebras with three fifth roots of plus unity and one fifth root of minus unity:

$$4 \quad \text{off} \quad a + b\sqrt[5]{+1} + c\sqrt[5]{+1} + d\sqrt[5]{+1} + e\sqrt[5]{-1} \tag{11.5}$$

Again we do not give the distance function.

There are six algebras with two fifth roots of plus unity and two fifth roots of minus unity:

$$6 \quad \text{off} \quad a + b\sqrt[5]{+1} + c\sqrt[5]{+1} + d\sqrt[5]{-1} + e\sqrt[5]{-1} \tag{11.6}$$

There are four algebras with one fifth root of plus unity and three fifth roots of minus unity:

$$4 \quad \text{off} \quad a + b\sqrt[5]{+1} + c\sqrt[5]{-1} + d\sqrt[5]{-1} + e\sqrt[5]{-1} \tag{11.7}$$

There is one algebra with four fifth roots of minus unity:

$$1 \quad \text{of} \quad a + b\sqrt[5]{-1} + c\sqrt[5]{-1} + d\sqrt[5]{-1} + e\sqrt[5]{-1} \tag{11.8}$$

The two single algebras, (11.3) & (11.8), are algebras, like the two 2-dimensional algebras and two of the 3-dimensional algebras, whose emergent distance functions are the same as the individual algebras. The existence of these single algebras is in stark contrast to the C_4 set of algebras which have no single algebras and therefore no emergent distance functions that are the same as the individual algebras.

We tabulate the algebras. We present for completeness the tabulated lesser order cyclic group algebras.

C_1	C_2	C_3	C_4	C_5
1	1 off $\left(1+\sqrt[2]{+1}\right)$	1 off $\left(1+2\sqrt[3]{+1}\right)$	4 off $\left(1+\sqrt[4]{+1}+2\sqrt[4]{+1}\right)$	1 off $\left(1+4\sqrt[5]{+1}\right)$
	1 off $\left(1+\sqrt[2]{-1}\right)$	2 off $\left(1+\sqrt[3]{+1}+\sqrt[3]{-1}\right)$	4 off $\left(1+\sqrt[4]{-1}+2\sqrt[4]{-1}\right)$	4 off $\left(1+3\sqrt[5]{+1}+\sqrt[5]{-1}\right)$
		1 off $\left(1+2\sqrt[3]{-1}\right)$		6 off $\left(1+2\sqrt[5]{+1}+2\sqrt[5]{-1}\right)$
				4 off $\left(1+\sqrt[5]{+1}+3\sqrt[5]{-1}\right)$
				1 off $\left(1+4\sqrt[5]{-1}\right)$

Prime order groups in general:
The numbers of different algebras within a prime order finite group are simply the numbers of ways that $(p-1)$ objects can be plus one or minus one; for example; there are six ways four objects can be arranged

with two plus ones and two minus ones. We expect this pattern to continue for all higher order cyclic groups of prime order.

The Order Six Cyclic Group

There are two finite groups of order six. These groups are the commutative cyclic group C_6 and the non-commutative symmetric group S_3. The group C_6 has two proper subgroups $\{C_2, C_3\}$.

The C_6 group:

The Standard form Cayley table of the C_6 group is:

$$C_6 = \begin{bmatrix} a & b & c & d & e & f \\ f & a & b & c & d & e \\ e & f & a & b & c & d \\ d & e & f & a & b & c \\ c & d & e & f & a & b \\ b & c & d & e & f & a \end{bmatrix} \qquad (12.1)$$

With parameters, this is the algebraic matrix form:

$$C_6 = \begin{bmatrix} a & b & c & d & e & f \\ P_{2,1}f & a & P_{2,3}b & P_{2,4}c & \dfrac{P_{2,4}P_{4,1}}{P_{3,1}}d & \dfrac{P_{2,3}P_{3,1}}{P_{2,1}}e \\ P_{3,1}e & \dfrac{P_{2,1}}{P_{2,3}}f & a & P_{2,4}b & \dfrac{(P_{2,4})^2 P_{4,1}}{P_{2,3}P_{3,1}}c & \dfrac{P_{2,4}P_{4,1}}{P_{2,1}}d \\ P_{4,1}d & \dfrac{P_{3,1}}{P_{2,4}}e & \dfrac{P_{2,1}}{P_{2,4}}f & a & \dfrac{P_{2,4}P_{4,1}}{P_{3,1}}b & \dfrac{P_{2,4}P_{4,1}}{P_{2,1}}c \\ P_{3,1}c & \dfrac{P_{3,1}}{P_{2,4}}d & \dfrac{P_{2,3}(P_{3,1})^2}{(P_{2,4})^2 P_{4,1}}e & \dfrac{P_{2,1}P_{3,1}}{P_{2,4}P_{4,1}}f & a & \dfrac{P_{2,3}P_{3,1}}{P_{2,1}}b \\ P_{2,1}b & \dfrac{P_{2,1}}{P_{2,3}}c & \dfrac{P_{2,1}}{P_{2,4}}d & \dfrac{P_{2,1}P_{3,1}}{P_{2,4}P_{4,1}}e & \dfrac{(P_{2,1})^2}{P_{2,3}P_{3,1}}f & a \end{bmatrix} \qquad (12.2)$$

This matrix, (12.2), is commutative, and all the algebras which it represents are commutative.

The division algebras are the exponentials of the algebraic matrix form, (12.2). There are $2^5 = 32$ separate division algebras corresponding to the 32 permutation of the five parameters as ± 1.

In this particular algebraic matrix form[18], the d variable is always a square root of plus or minus unity and the $\{c, e\}$ variables are always the cube roots of plus or minus unity. The other variables are the sixth roots of plus or minus unity. This reflects the subgroup structure of the C_6 group expressed as the orders of its elements.

There are eight sets of algebras:

$$a + b\sqrt[6]{-1} + c\sqrt[3]{-1} + d\sqrt[2]{-1} + e\sqrt[3]{-1} + f\sqrt[6]{-1} \qquad 4 \; \textit{off} \qquad (12.3)$$

$$a + b\sqrt[6]{+1} + c\sqrt[3]{+1} + d\sqrt[2]{+1} + e\sqrt[3]{+1} + f\sqrt[6]{+1} \qquad 4 \; \textit{off} \qquad (12.4)$$

$$a + b\sqrt[6]{+1} + c\sqrt[3]{-1} + d\sqrt[2]{+1} + e\sqrt[3]{+1} + f\sqrt[6]{+1} \qquad 4 \; \textit{off} \qquad (12.5)$$

$$a + b\sqrt[6]{+1} + c\sqrt[3]{+1} + d\sqrt[2]{+1} + e\sqrt[3]{-1} + f\sqrt[6]{+1} \qquad 4 \; \textit{off} \qquad (12.6)$$

$$a + b\sqrt[6]{-1} + c\sqrt[3]{+1} + d\sqrt[2]{-1} + e\sqrt[3]{-1} + f\sqrt[6]{-1} \qquad 4 \; \textit{off} \qquad (12.7)$$

$$a + b\sqrt[6]{-1} + c\sqrt[3]{-1} + d\sqrt[2]{-1} + e\sqrt[3]{+1} + f\sqrt[6]{-1} \qquad 4 \; \textit{off} \qquad (12.8)$$

$$a + b\sqrt[6]{-1} + c\sqrt[3]{+1} + d\sqrt[2]{-1} + e\sqrt[3]{+1} + f\sqrt[6]{-1} \qquad 4 \; \textit{off} \qquad (12.9)$$

$$a + b\sqrt[6]{+1} + c\sqrt[3]{-1} + d\sqrt[2]{+1} + e\sqrt[3]{-1} + f\sqrt[6]{+1} \qquad 4 \; \textit{off} \qquad (12.10)$$

However, of these (12.5) & (12.6) are the same algebra with a swapped variable and (12.7) & (12.8) are the same algebra with a swapped variable. The list of algebras is more correctly presented as the six non-isomorphic algebras:

$$a + b\sqrt[6]{-1} + c\sqrt[3]{-1} + d\sqrt[2]{-1} + e\sqrt[3]{-1} + f\sqrt[6]{-1} \qquad 4 \; \textit{off} \qquad (12.11)$$

$$a + b\sqrt[6]{+1} + c\sqrt[3]{+1} + d\sqrt[2]{+1} + e\sqrt[3]{+1} + f\sqrt[6]{+1} \qquad 4 \; \textit{off} \qquad (12.12)$$

$$a + b\sqrt[6]{+1} + c\sqrt[3]{+1} + d\sqrt[2]{+1} + e\sqrt[3]{-1} + f\sqrt[6]{+1} \qquad 8 \; \textit{off} \qquad (12.13)$$

$$a + b\sqrt[6]{-1} + c\sqrt[3]{+1} + d\sqrt[2]{-1} + e\sqrt[3]{-1} + f\sqrt[6]{-1} \qquad 8 \; \textit{off} \qquad (12.14)$$

$$a + b\sqrt[6]{-1} + c\sqrt[3]{+1} + d\sqrt[2]{-1} + e\sqrt[3]{+1} + f\sqrt[6]{-1} \qquad 4 \; \textit{off} \qquad (12.15)$$

$$a + b\sqrt[6]{+1} + c\sqrt[3]{-1} + d\sqrt[2]{+1} + e\sqrt[3]{-1} + f\sqrt[6]{+1} \qquad 4 \; \textit{off} \qquad (12.16)$$

Looking back at the computer program, we see that we took account of this.

---

[18] The reader is reminded that the Standard form of the Cayley table is not unique and so the algebraic matrix form is not unique. In the same way that all forms of a Cayley table describe the same group, so all forms of algebraic matrix form describe the same algebras.

The C_6 emergent distance functions:

Although the individual distance functions will factorise, they do not factorise to a quadratic form except for pairs of variables. Hence there are no emergent distance functions which hold geometric spaces.

Summary C_6:

There are no emergent distance functions which support a full set of rotations. Thus, there is no geometric space supported by the C_6 group.

$$C_6$$

$$\text{4 off } 1+\sqrt[2]{+1}+2\sqrt[3]{+1}+2\sqrt[6]{+1}$$

$$\text{4 off } 1+\sqrt[2]{-1}+2\sqrt[3]{-1}+2\sqrt[6]{-1}$$

$$\text{8 off } 1+\sqrt[2]{+1}+\sqrt[3]{-1}+\sqrt[3]{+1}+2\sqrt[6]{+1}$$

$$\text{8 off } 1+\sqrt[2]{-1}+\sqrt[3]{-1}+\sqrt[3]{+1}+2\sqrt[6]{-1}$$

$$\text{4 off } 1+\sqrt[2]{-1}+2\sqrt[3]{+1}+2\sqrt[6]{-1}$$

$$\text{4 off } 1+\sqrt[2]{+1}+2\sqrt[3]{-1}+2\sqrt[6]{-1}$$

The Order Seven Cyclic group

Since seven is a prime number, there is only one order seven group, C_7. As with all cyclic groups, this is a commutative group. As with all groups of prime order, this group has no proper subgroups. As with all cyclic groups, the Standard form Cayley table can be written down with variables parallel to the leading diagonal. C_7 has Standard form Cayley table:

$$C_7 = \begin{bmatrix} a & b & c & d & e & f & g \\ g & a & b & c & d & e & f \\ f & g & a & b & c & d & e \\ e & f & g & a & b & c & d \\ d & e & f & g & a & b & c \\ c & d & e & f & g & a & b \\ b & c & d & e & f & g & a \end{bmatrix} \tag{13.1}$$

The algebraic matrix form is:

$$C_7 = \begin{bmatrix} a & b & c & d & e & f & g \\ P_{2,1}g & a & P_{2,3}b & P_{2,4}c & P_{2,5}d & \dfrac{P_{2,4}P_{4,1}}{P_{3,1}}e & \dfrac{P_{2,3}P_{3,1}}{P_{2,1}}f \\ P_{3,1}f & \dfrac{P_{2,1}}{P_{2,3}}g & a & P_{2,4}b & \dfrac{P_{2,4}P_{2,5}}{P_{2,3}}c & \dfrac{P_{2,4}P_{2,5}P_{4,1}}{P_{2,3}P_{3,1}}d & \dfrac{P_{2,4}P_{4,1}}{P_{2,1}}e \\ P_{4,1}e & \dfrac{P_{3,1}}{P_{2,4}}f & \dfrac{P_{2,1}}{P_{2,4}}g & a & P_{2,5}b & \dfrac{P_{2,4}P_{2,5}P_{4,1}}{P_{2,3}P_{3,1}}c & \dfrac{P_{2,5}P_{4,1}}{P_{2,1}}d \\ P_{4,1}d & \dfrac{P_{4,1}}{P_{2,5}}e & \dfrac{P_{2,3}P_{3,1}}{P_{2,4}P_{2,5}}f & \dfrac{P_{2,1}}{P_{2,5}}g & a & \dfrac{P_{2,4}P_{4,1}}{P_{3,1}}b & \dfrac{P_{2,4}P_{4,1}}{P_{2,1}}c \\ P_{3,1}c & \dfrac{P_{2,1}}{P_{2,4}}d & \dfrac{P_{2,3}P_{3,1}}{P_{2,4}P_{2,5}}e & \dfrac{P_{2,3}\left(P_{3,1}\right)^2}{P_{2,4}P_{2,5}P_{4,1}}f & \dfrac{P_{2,1}P_{3,1}}{P_{2,4}P_{4,1}}g & a & \dfrac{P_{2,3}P_{3,1}}{P_{2,1}}b \\ P_{2,1}b & \dfrac{P_{2,1}}{P_{2,3}}c & \dfrac{P_{2,1}}{P_{2,4}}d & \dfrac{P_{2,1}}{P_{2,5}}e & \dfrac{P_{2,3}P_{3,1}}{P_{2,4}P_{4,1}}f & \dfrac{\left(P_{2,1}\right)^2}{P_{2,3}P_{3,1}}g & a \end{bmatrix} \tag{13.2}$$

This matrix, (13.2), is commutative, and all the algebras which it represents are commutative.

There are a total of 64 division algebras (2^6 permutations of the signs of six parameters). Of these 64 division algebras there are seven algebraically non-isomorphic algebras which occur in the following numbers:

$$a + b\sqrt[7]{+1} + c\sqrt[7]{+1} + d\sqrt[7]{+1} + e\sqrt[7]{+1} + f\sqrt[7]{+1} + g\sqrt[7]{+1} \qquad 1 \;\; \textit{off} \qquad (13.3)$$

$$a + b\sqrt[7]{+1} + c\sqrt[7]{+1} + d\sqrt[7]{+1} + e\sqrt[7]{+1} + f\sqrt[7]{+1} + g\sqrt[7]{-1} \qquad 6 \;\; \textit{off} \qquad (13.4)$$

$$a + b\sqrt[7]{+1} + c\sqrt[7]{+1} + d\sqrt[7]{+1} + e\sqrt[7]{+1} + f\sqrt[7]{-1} + g\sqrt[7]{-1} \qquad 15 \;\; \textit{off} \qquad (13.5)$$

$$a + b\sqrt[7]{+1} + c\sqrt[7]{+1} + d\sqrt[7]{+1} + e\sqrt[7]{-1} + f\sqrt[7]{-1} + g\sqrt[7]{-1} \qquad 20 \;\; \textit{off} \qquad (13.6)$$

$$a + b\sqrt[7]{+1} + c\sqrt[7]{+1} + d\sqrt[7]{-1} + e\sqrt[7]{-1} + f\sqrt[7]{-1} + g\sqrt[7]{-1} \qquad 15 \;\; \textit{off} \qquad (13.7)$$

$$a + b\sqrt[7]{+1} + c\sqrt[7]{-1} + d\sqrt[7]{-1} + e\sqrt[7]{-1} + f\sqrt[7]{-1} + g\sqrt[7]{-1} \qquad 6 \;\; \textit{off} \qquad (13.8)$$

$$a + b\sqrt[7]{-1} + c\sqrt[7]{-1} + d\sqrt[7]{-1} + e\sqrt[7]{-1} + f\sqrt[7]{-1} + g\sqrt[7]{-1} \qquad 1 \;\; \textit{off} \qquad (13.9)$$

Comparing these numbers of isomorphic algebras with the same from the lesser order prime cyclic groups, we see a pattern. There are always the same number of isomorphic algebras as the order of the prime cyclic group and they are symmetrically balanced in terms of the numbers of nth roots of plus and minus unity.

Since C_7 is a prime cyclic group, the emergent distance functions are of the form d^{prime} and cannot be factored into distance functions that support lesser dimensional rotations.

$$C_7$$

$$1 \text{ off } 1 + 6\sqrt[7]{+1}$$

$$6 \text{ off } 1 + 5\sqrt[7]{+1} + \sqrt[7]{-1}$$

$$15 \text{ off } 1 + 4\sqrt[7]{+1} + 2\sqrt[7]{-1}$$

$$20 \text{ off } 1 + 3\sqrt[7]{+1} + 3\sqrt[7]{-1}$$

$$15 \text{ off } 1 + 2\sqrt[7]{+1} + 4\sqrt[7]{-1}$$

$$6 \text{ off } 1 + \sqrt[7]{+1} + 5\sqrt[7]{-1}$$

$$1 \text{ off } 1 + 6\sqrt[7]{-1}$$

The Order Eight Cyclic Group

There are five order eight groups. These are the three commutative groups C_8 and $C_2 \times C_4$ and $C_2 \times C_2 \times C_2$ and the two non-commutative groups D_4 and Q.

The C_8 group:

The commutative C_8 group has one C_2 subgroup and one C_4 subgroup. Being a cyclic group, we can write its Standard form Cayley table as variables running parallel to the leading diagonal:

$$C_8 = \begin{bmatrix} a & b & c & d & e & f & g & h \\ h & a & b & c & d & e & f & g \\ g & h & a & b & c & d & e & f \\ f & g & h & a & b & c & d & e \\ e & f & g & h & a & b & c & d \\ d & e & f & g & h & a & b & c \\ c & d & e & f & g & h & a & b \\ b & c & d & e & f & g & h & a \end{bmatrix} \tag{14.1}$$

Because the C_8 group has a C_4 subgroup, we know that it cannot hold any emergent distance functions which form geometric spaces because the C_4 group has no geometric spaces. The C_8 finite group has algebraic matrix form:

$$C_8 = \begin{bmatrix}
a & b & c & d & e & f & g & h \\[4pt]
P_{2,1}h & a & P_{2,3}b & P_{2,4}c & P_{2,5}d & \dfrac{P_{2,5}P_{5,1}}{P_{4,1}}e & \dfrac{P_{2,4}P_{4,1}}{P_{3,1}}f & \dfrac{P_{2,3}P_{3,1}}{P_{2,1}}g \\[12pt]
P_{3,1}g & \dfrac{P_{2,1}}{P_{2,3}}h & a & P_{2,4}b & \dfrac{P_{2,4}P_{2,5}}{P_{2,3}}c & \dfrac{\left(P_{2,5}\right)^2 P_{5,1}}{P_{2,3}P_{4,1}}d & \dfrac{P_{2,4}P_{2,5}P_{5,1}}{P_{2,3}P_{3,1}}e & \dfrac{P_{2,4}P_{4,1}}{P_{2,1}}f \\[12pt]
P_{4,1}f & \dfrac{P_{3,1}}{P_{2,4}}g & \dfrac{P_{2,1}}{P_{2,4}}h & a & P_{2,5}b & \dfrac{\left(P_{2,5}\right)^2 P_{5,1}}{P_{2,3}P_{4,1}}c & \dfrac{\left(P_{2,5}\right)^2 P_{5,1}}{P_{2,3}P_{3,1}}d & \dfrac{P_{2,5}P_{5,1}}{P_{2,1}}e \\[12pt]
P_{5,1}e & \dfrac{P_{4,1}}{P_{2,5}}f & \dfrac{P_{2,3}P_{3,1}}{P_{2,4}P_{2,5}}g & \dfrac{P_{2,1}}{P_{2,5}}h & a & \dfrac{P_{2,5}P_{5,1}}{P_{4,1}}b & \dfrac{P_{2,4}P_{2,5}P_{5,1}}{P_{2,3}P_{3,1}}c & \dfrac{P_{2,5}P_{5,1}}{P_{2,1}}d \\[12pt]
P_{4,1}d & \dfrac{P_{4,1}}{P_{2,5}}e & \dfrac{P_{2,3}\left(P_{4,1}\right)^2}{\left(P_{2,5}\right)^2 P_{5,1}}f & \dfrac{P_{2,3}P_{3,1}P_{4,1}}{\left(P_{2,5}\right)^2 P_{5,1}}g & \dfrac{P_{2,1}P_{4,1}}{P_{2,5}P_{5,1}}h & a & \dfrac{P_{2,4}P_{4,1}}{P_{3,1}}b & \dfrac{P_{2,4}P_{4,1}}{P_{2,1}}c \\[12pt]
P_{3,1}c & \dfrac{P_{3,1}}{P_{2,4}}d & \dfrac{P_{2,3}P_{3,1}}{P_{2,4}P_{2,5}}e & \dfrac{P_{2,3}P_{3,1}P_{4,1}}{\left(P_{2,5}\right)^2 P_{5,1}}f & \dfrac{P_{2,3}\left(P_{3,1}\right)^2}{P_{2,4}P_{2,5}P_{5,1}}g & \dfrac{P_{2,1}P_{3,1}}{P_{2,4}P_{4,1}}h & a & \dfrac{P_{2,3}P_{3,1}}{P_{2,1}}b \\[12pt]
P_{2,1}b & \dfrac{P_{2,1}}{P_{2,3}}c & \dfrac{P_{2,1}}{P_{2,4}}d & \dfrac{P_{2,1}}{P_{2,5}}e & \dfrac{P_{2,1}P_{4,1}}{P_{2,5}P_{5,1}}f & \dfrac{P_{2,1}P_{3,1}}{P_{2,4}P_{4,1}}g & \dfrac{\left(P_{2,1}\right)^2}{P_{2,3}P_{3,1}}h & a
\end{bmatrix}$$

(14.2)

There are 128 division algebras given by the permutations of the parameters as ±1. These algebras are of only two algebraically non-isomorphic forms:

$$a + b\sqrt[8]{+1} + c\sqrt[4]{+1} + d\sqrt[8]{+1} + e\sqrt[2]{+1} + f\sqrt[8]{+1} + g\sqrt[4]{+1} + h\sqrt[8]{+1} \qquad 64 \; \textit{off} \qquad (14.3)$$

$$a + b\sqrt[8]{-1} + c\sqrt[4]{-1} + d\sqrt[8]{-1} + e\sqrt[2]{-1} + f\sqrt[8]{-1} + g\sqrt[4]{-1} + h\sqrt[8]{-1} \qquad 64 \; \textit{off} \qquad (14.4)$$

Other Standard form Cayley tables 'swap' the variables which are the square roots, 4th roots and 8th roots.

Emergent distance functions in C_8 :

The C_8 emergent distance functions cannot support 2-dimensional rotations.

The Order Nine Cyclic Group

There are two order nine groups, the cyclic group C_9 and the group $C_3 \times C_3$. The cyclic group C_9 has one C_3 subgroup.

The Standard form Cayley table of C_9 is:

$$C_9 = \begin{bmatrix} a & b & c & d & e & f & g & h & i \\ i & a & b & c & d & e & f & g & h \\ h & i & a & b & c & d & e & f & g \\ g & h & i & a & b & c & d & e & f \\ f & g & h & i & a & b & c & d & e \\ e & f & g & h & i & a & b & c & d \\ d & e & f & g & h & i & a & b & c \\ c & d & e & f & g & h & i & a & b \\ b & c & d & e & f & g & h & i & a \end{bmatrix} \qquad (15.1)$$

The C_9 algebraic matrix form is:

$$C_9 =$$

$$
\begin{bmatrix}
a & b & c & d & e & f & g & h & i \\[6pt]
P_{2,1}i & a & P_{2,3}b & P_{2,4}c & P_{2,5}d & P_{2,6}e & \dfrac{P_{2,5}P_{5,1}}{P_{4,1}}f & \dfrac{P_{2,4}P_{4,1}}{P_{3,1}}g & \dfrac{P_{2,3}P_{3,1}}{P_{2,1}}h \\[12pt]
P_{3,1}h & \dfrac{P_{2,1}}{P_{2,3}}i & a & P_{2,4}b & \dfrac{P_{2,4}P_{2,5}}{P_{2,3}}c & \dfrac{P_{2,5}P_{2,6}}{P_{2,3}}d & \dfrac{P_{2,5}P_{2,6}P_{5,1}}{P_{2,3}P_{4,1}}e & \dfrac{P_{2,4}P_{2,5}P_{5,1}}{P_{2,3}P_{3,1}}f & \dfrac{P_{2,4}P_{4,1}}{P_{2,1}}g \\[12pt]
P_{4,1}g & \dfrac{P_{3,1}}{P_{2,4}}h & \dfrac{P_{2,1}}{P_{2,4}}i & a & P_{2,5}b & \dfrac{P_{2,5}P_{2,6}}{P_{2,3}}c & \dfrac{P_{2,5}P_{2,5}P_{2,6}P_{5,1}}{P_{2,3}P_{2,4}P_{4,1}}d & \dfrac{P_{2,5}P_{2,6}P_{5,1}}{P_{2,3}P_{3,1}}e & \dfrac{P_{2,5}P_{5,1}}{P_{2,1}}f \\[12pt]
P_{5,1}f & \dfrac{P_{4,1}}{P_{2,5}}g & \dfrac{P_{2,3}P_{3,1}}{P_{2,4}P_{2,5}}h & \dfrac{P_{2,1}}{P_{2,5}}i & a & P_{2,6}b & \dfrac{P_{2,5}P_{2,6}P_{5,1}}{P_{2,3}P_{4,1}}c & \dfrac{P_{2,5}P_{2,6}P_{5,1}}{P_{2,3}P_{3,1}}d & \dfrac{P_{2,6}P_{5,1}}{P_{2,1}}e \\[12pt]
P_{5,1}e & \dfrac{P_{5,1}}{P_{2,6}}f & \dfrac{P_{2,3}P_{4,1}}{P_{2,5}P_{2,6}}g & \dfrac{P_{2,3}P_{3,1}}{P_{2,5}P_{2,6}}h & \dfrac{P_{2,1}}{P_{2,6}}i & a & \dfrac{P_{2,5}P_{5,1}}{P_{4,1}}b & \dfrac{P_{2,4}P_{2,5}P_{5,1}}{P_{2,3}P_{3,1}}c & \dfrac{P_{2,5}P_{5,1}}{P_{2,1}}d \\[12pt]
P_{4,1}d & \dfrac{P_{4,1}}{P_{2,5}}e & \dfrac{P_{2,3}P_{4,1}}{P_{2,5}P_{2,6}}f & \dfrac{P_{2,3}P_{2,4}P_{4,1}P_{4,1}}{P_{2,5}P_{2,5}P_{2,6}P_{5,1}}g & \dfrac{P_{2,3}P_{3,1}P_{4,1}}{P_{2,5}P_{2,6}P_{5,1}}h & \dfrac{P_{2,1}P_{4,1}}{P_{2,5}P_{5,1}}i & a & \dfrac{P_{2,4}P_{4,1}}{P_{3,1}}b & \dfrac{P_{2,4}P_{4,1}}{P_{2,1}}c \\[12pt]
P_{3,1}c & \dfrac{P_{3,1}}{P_{2,4}}d & \dfrac{P_{2,3}P_{3,1}}{P_{2,4}P_{2,5}}e & \dfrac{P_{2,3}P_{3,1}}{P_{2,5}P_{2,6}}f & \dfrac{P_{2,3}P_{3,1}P_{4,1}}{P_{2,5}P_{2,6}P_{5,1}}g & \dfrac{P_{2,3}P_{3,1}P_{3,1}}{P_{2,4}P_{2,5}P_{5,1}}h & \dfrac{P_{2,1}P_{3,1}}{P_{2,4}P_{4,1}}i & a & \dfrac{P_{2,3}P_{3,1}}{P_{2,1}}b \\[12pt]
P_{2,1}b & \dfrac{P_{2,1}}{P_{2,3}}c & \dfrac{P_{2,1}}{P_{2,4}}d & \dfrac{P_{2,1}}{P_{2,5}}e & \dfrac{P_{2,1}}{P_{2,6}}f & \dfrac{P_{2,1}P_{4,1}}{P_{2,5}P_{5,1}}g & \dfrac{P_{2,1}P_{3,1}}{P_{2,4}P_{4,1}}h & \dfrac{P_{2,1}P_{2,1}}{P_{2,3}P_{3,1}}i & a
\end{bmatrix}
$$

$$(15.2)$$

There are 256 separate division algebras within the finite group C_9. The general form of these algebras with the Standard form Cayley table used for cyclic groups in this book is:

$$a + b\sqrt[9]{\pm 1} + c\sqrt[9]{\pm 1} + d\sqrt[3]{\pm 1} + e\sqrt[9]{\pm 1} + f\sqrt[9]{\pm 1} + g\sqrt[3]{\pm 1} + h\sqrt[9]{\pm 1} + i\sqrt[9]{\pm 1} \tag{15.3}$$

These algebras are of twenty one different types:

$$
\begin{array}{lll}
1 & \text{of} & a + 2\sqrt[3]{+1} + 6\sqrt[9]{+1} \\
1 & \text{of} & a + 2\sqrt[3]{-1} + 6\sqrt[9]{-1}
\end{array} \tag{15.4}
$$

$$
\begin{array}{lll}
1 & \text{of} & a + 2\sqrt[3]{-1} + 6\sqrt[9]{+1} \\
1 & \text{of} & a + 2\sqrt[3]{+1} + 6\sqrt[9]{-1}
\end{array} \tag{15.5}
$$

$$
\begin{array}{lll}
2 & \text{off} & a + \sqrt[3]{+1} + \sqrt[3]{-1} + 6\sqrt[9]{+1} \\
2 & \text{off} & a + \sqrt[3]{+1} + \sqrt[3]{-1} + 6\sqrt[9]{-1}
\end{array} \tag{15.6}
$$

$$
\begin{array}{lll}
6 & \text{off} & a + 2\sqrt[3]{+1} + 5\sqrt[3]{+1} + \sqrt[9]{-1} \\
6 & \text{off} & a + 2\sqrt[3]{-1} + 5\sqrt[3]{-1} + \sqrt[9]{+1}
\end{array} \tag{15.7}
$$

$$6 \quad \text{off} \quad a + 2\sqrt[3]{+1} + 5\sqrt[3]{-1} + \sqrt[9]{+1}$$
$$6 \quad \text{off} \quad a + 2\sqrt[3]{-1} + 5\sqrt[3]{+1} + \sqrt[9]{-1} \tag{15.8}$$

$$12 \quad \text{off} \quad a + \sqrt[3]{+1} + \sqrt[3]{-1} + 5\sqrt[9]{+1} + \sqrt[9]{-1}$$
$$12 \quad \text{off} \quad a + \sqrt[3]{-1} + \sqrt[3]{+1} + 5\sqrt[9]{-1} + \sqrt[9]{+1} \tag{15.9}$$

$$15 \quad \text{off} \quad a + 2\sqrt[3]{-1} + 4\sqrt[9]{-1} + 2\sqrt[9]{+1}$$
$$15 \quad \text{off} \quad a + 2\sqrt[3]{+1} + 4\sqrt[9]{+1} + 2\sqrt[9]{-1} \tag{15.10}$$

$$15 \quad \text{off} \quad a + 2\sqrt[3]{-1} + 4\sqrt[9]{+1} + 2\sqrt[9]{-1}$$
$$15 \quad \text{off} \quad a + 2\sqrt[3]{+1} + 4\sqrt[9]{-1} + 2\sqrt[9]{+1} \tag{15.11}$$

$$20 \quad \text{off} \quad a + 2\sqrt[3]{-1} + 3\sqrt[9]{+1} + 3\sqrt[9]{-1}$$
$$20 \quad \text{off} \quad a + 2\sqrt[3]{+1} + 3\sqrt[9]{-1} + 3\sqrt[9]{+1} \tag{15.12}$$

$$30 \quad \text{off} \quad a + \sqrt[3]{+1} + \sqrt[3]{-1} + 4\sqrt[9]{+1} + 2\sqrt[9]{-1}$$
$$30 \quad \text{off} \quad a + \sqrt[3]{-1} + \sqrt[3]{+1} + 4\sqrt[9]{-1} + 2\sqrt[9]{+1} \tag{15.13}$$

$$40 \quad \text{off} \quad a + \sqrt[3]{+1} + \sqrt[3]{-1} + 3\sqrt[9]{+1} + 3\sqrt[9]{-1} \tag{15.14}$$

It is quite a surprise to find twenty one different algebras within this group. We reiterate that we do not clearly understand the distribution of the division algebras over the finite groups.

Emergent distance functions of C_9:

Since nine is an odd number, the group C_9 cannot support geometric spaces that have 2-dimensional rotations.

Computer calculations show that C_9 cannot support geometric spaces that have 3-dimensional rotations.

The Higher Order Cyclic Groups

The nature of the Standard form Cayley tables of the Cyclic groups adopted by this book is such that the variables occur in lines parallel to the leading diagonal. There are many examples previous to this chapter. Since the Standard form Cayley table is easily written following the many examples in earlier chapters, we will not present the Standard form Cayley tables of the higher order cyclic groups.

In an earlier chapter, we presented a computer program for calculating the algebraic matrix forms of the cyclic groups. The program is not perfect, and, for higher order groups, we have to do some elimination calculations by hand. We have not room to present the algebraic matrix forms of the higher order cyclic groups. If needed, they can be calculated by the computer program previously given. Indeed, the only sensible way to handle them seems to be by computer.

The order 10 cyclic group:

The order ten finite group C_{10} has two subgroups, $\{C_2, C_5\}$, There are $2^{(10-1)} = 512$ separate division algebras within the C_{10} group. The general form of these algebras with the Standard form Cayley table used for cyclic groups in this book is (remember we need to take the exponential of the algebraic matrix form):

$$a + b\sqrt[10]{\pm 1} + c\sqrt[5]{\pm 1} + d\sqrt[10]{\pm 1} + e\sqrt[5]{\pm 1} + f\sqrt[2]{\pm 1} + g\sqrt[5]{\pm 1} + h\sqrt[10]{\pm 1} + i\sqrt[5]{\pm 1} + j\sqrt[10]{\pm 1} \tag{16.1}$$

There are ten non-isomorphic algebras:

$$
\begin{array}{lll}
16 & \text{off} & a + \sqrt[2]{+1} + 4\sqrt[5]{+1} + 4\sqrt[10]{+1} \\
16 & \text{off} & a + \sqrt[2]{-1} + 4\sqrt[5]{-1} + 4\sqrt[10]{-1}
\end{array}
\tag{16.2}
$$

$$
\begin{array}{lll}
16 & \text{off} & a + \sqrt[2]{+1} + 4\sqrt[5]{-1} + 4\sqrt[10]{+1} \\
16 & \text{off} & a + \sqrt[2]{-1} + 4\sqrt[5]{+1} + 4\sqrt[10]{-1}
\end{array}
\tag{16.3}
$$

$$
\begin{array}{lll}
64 & \text{off} & a + \sqrt[2]{+1} + \sqrt[5]{-1} + 3\sqrt[5]{+1} + 4\sqrt[10]{+1} \\
64 & \text{off} & a + \sqrt[2]{-1} + \sqrt[5]{+1} + 3\sqrt[5]{-1} + 4\sqrt[10]{-1}
\end{array}
\tag{16.4}
$$

$$
\begin{array}{lll}
64 & \text{off} & a + \sqrt[2]{+1} + \sqrt[5]{+1} + 3\sqrt[5]{-1} + 4\sqrt[10]{+1} \\
64 & \text{off} & a + \sqrt[2]{-1} + \sqrt[5]{-1} + 3\sqrt[5]{+1} + 4\sqrt[10]{-1}
\end{array}
\tag{16.5}
$$

$$
\begin{array}{lll}
96 & \text{off} & a + \sqrt[2]{+1} + 2\sqrt[5]{+1} + 2\sqrt[5]{-1} + 4\sqrt[10]{+1} \\
96 & \text{off} & a + \sqrt[2]{-1} + 2\sqrt[5]{-1} + 2\sqrt[5]{+1} + 4\sqrt[10]{-1}
\end{array}
\tag{16.6}
$$

Computer calculation shows that there are no geometric spaces within the C_{10} group. Since C_{10} contains a subgroup of prime order, C_5, it cannot contain a geometric space. Recall that since groups of prime order, other than two in the 2-dimensional case or three in the 3-dimensional case etc., cannot be reduced to the quadratic form, any larger group that has a subgroup of prime order cannot hold a geometric space because it must reduce when all the variables in the group other than those that form the prime order subgroup are zero.

The order 11 cyclic group:

Eleven is a prime number and so the finite cyclic group C_{11} has no subgroups. There are $2^{(11-1)} = 1024$ separate division algebras within the C_{11} group. The general form of these algebras with the Standard form Cayley table used for cyclic groups in this book is:

$$a + b\sqrt[11]{\pm 1} + c\sqrt[11]{\pm 1} + d\sqrt[11]{\pm 1} + e\sqrt[11]{\pm 1} + f\sqrt[11]{\pm 1} + g\sqrt[11]{\pm 1} + h\sqrt[11]{\pm 1} + i\sqrt[11]{\pm 1} + j\sqrt[11]{\pm 1} + k\sqrt[11]{\pm 1} \qquad (16.7)$$

These algebras are of eleven different types:

$$
\begin{array}{lll}
1 & \text{of} & 1 + 10\sqrt[11]{+1} \\
1 & \text{of} & 1 + 10\sqrt[11]{-1}
\end{array} \qquad (16.8)
$$

$$
\begin{array}{lll}
10 & \text{off} & 1 + 9\sqrt[11]{+1} + \sqrt[11]{-1} \\
10 & \text{off} & 1 + \sqrt[11]{+1} + 9\sqrt[11]{-1}
\end{array} \qquad (16.9)
$$

$$
\begin{array}{lll}
45 & \text{off} & 1 + 8\sqrt[11]{+1} + 2\sqrt[11]{-1} \\
45 & \text{off} & 1 + 2\sqrt[11]{+1} + 8\sqrt[11]{-1}
\end{array} \qquad (16.10)
$$

$$
\begin{array}{lll}
120 & \text{off} & 1 + 7\sqrt[11]{+1} + 3\sqrt[11]{-1} \\
120 & \text{off} & 1 + 3\sqrt[11]{+1} + 7\sqrt[11]{-1}
\end{array} \qquad (16.11)
$$

$$
\begin{array}{lll}
210 & \text{off} & 1 + 6\sqrt[11]{+1} + 4\sqrt[11]{-1} \\
210 & \text{off} & 1 + 4\sqrt[11]{+1} + 6\sqrt[11]{-1}
\end{array} \qquad (16.12)
$$

$$
\begin{array}{lll}
252 & \text{off} & 1 + 5\sqrt[11]{+1} + 5\sqrt[11]{-1}
\end{array} \qquad (16.13)
$$

Being of prime order there are no geometric spaces within the C_{11} group.

The order 12 cyclic group:

The order twelve cyclic finite group, C_{12}, has one copy of each of the subgroups $\{C_2, C_3, C_4, C_6\}$. The order twelve cyclic group has 2,048 separate algebras. With the Standard form Cayley table used in this book, the form of these algebras is:

$$a + b\sqrt[12]{\pm 1} + c\sqrt[6]{\pm 1} + d\sqrt[4]{\pm 1} + e\sqrt[3]{\pm 1} + f\sqrt[12]{\pm 1} + g\sqrt[2]{\pm 1} + h\sqrt[12]{\pm 1} + i\sqrt[3]{\pm 1} + j\sqrt[4]{\pm 1} + k\sqrt[6]{\pm 1} + l\sqrt[12]{\pm 1} \qquad (16.14)$$

There are six different (non-isomorphic) division algebras. They are of the form:

$$
\begin{aligned}
256 \quad \text{off} \quad & 1+\sqrt[2]{+1}+2\sqrt[3]{+1}+2\sqrt[4]{+1}+2\sqrt[6]{+1}+4\sqrt[12]{+1} \\
256 \quad \text{off} \quad & 1+\sqrt[2]{-1}+2\sqrt[3]{-1}+2\sqrt[4]{-1}+2\sqrt[6]{-1}+4\sqrt[12]{-1}
\end{aligned}
\tag{16.15}
$$

$$
\begin{aligned}
256 \quad \text{off} \quad & 1+\sqrt[2]{+1}+2\sqrt[3]{-1}+2\sqrt[4]{+1}+2\sqrt[6]{+1}+4\sqrt[12]{+1} \\
256 \quad \text{off} \quad & 1+\sqrt[2]{-1}+2\sqrt[3]{+1}+2\sqrt[4]{-1}+2\sqrt[6]{-1}+4\sqrt[12]{-1}
\end{aligned}
\tag{16.16}
$$

$$
\begin{aligned}
512 \quad \text{off} \quad & 1+\sqrt[2]{+1}+\sqrt[3]{+1}+\sqrt[3]{-1}+2\sqrt[4]{+1}+2\sqrt[6]{+1}+4\sqrt[12]{+1} \\
512 \quad \text{off} \quad & 1+\sqrt[2]{-1}+\sqrt[3]{-1}+\sqrt[3]{+1}+2\sqrt[4]{-1}+2\sqrt[6]{-1}+4\sqrt[12]{-1}
\end{aligned}
\tag{16.17}
$$

Since the C_{12} finite group has a C_4 subgroup (the same applies to the C_3 & C_6 subgroups), there can be no geometric spaces in this group.

The order 13 cyclic group:

The order thirteen cyclic finite group, C_{13}, being of prime order has no subgroups. The order thirteen cyclic group has 4,096 separate algebras. With the Standard form Cayley table used in this book, the form of these algebras is:

$$
a+b\sqrt[13]{\pm1}+c\sqrt[13]{\pm1}+d\sqrt[13]{\pm1}+e\sqrt[13]{\pm1}+f\sqrt[13]{\pm1}+g\sqrt[13]{\pm1}+h\sqrt[13]{\pm1}+i\sqrt[13]{\pm1}+j\sqrt[13]{\pm1}+k\sqrt[13]{\pm1}+l\sqrt[13]{\pm1}+m\sqrt[13]{\pm1}
\tag{16.18}
$$

There are thirteen different (non-isomorphic) division algebras. They are of the form:

$$
\begin{aligned}
1 \quad \text{of} \quad & 1+12\sqrt[13]{+1} \\
1 \quad \text{of} \quad & 1+12\sqrt[13]{-1}
\end{aligned}
\tag{16.19}
$$

$$
\begin{aligned}
12 \quad \text{off} \quad & 1+11\sqrt[13]{+1}+\sqrt[13]{-1} \\
12 \quad \text{off} \quad & 1+11\sqrt[13]{-1}+\sqrt[13]{+1}
\end{aligned}
\tag{16.20}
$$

$$
\begin{aligned}
66 \quad \text{off} \quad & 1+10\sqrt[13]{+1}+2\sqrt[13]{-1} \\
66 \quad \text{off} \quad & 1+10\sqrt[13]{-1}+2\sqrt[13]{+1}
\end{aligned}
\tag{16.21}
$$

$$
\begin{aligned}
220 \quad \text{off} \quad & 1+9\sqrt[13]{+1}+3\sqrt[13]{-1} \\
220 \quad \text{off} \quad & 1+9\sqrt[13]{-1}+3\sqrt[13]{+1}
\end{aligned}
\tag{16.22}
$$

$$
\begin{aligned}
495 \quad \text{off} \quad & 1+8\sqrt[13]{+1}+4\sqrt[13]{-1} \\
495 \quad \text{off} \quad & 1+8\sqrt[13]{-1}+4\sqrt[13]{+1}
\end{aligned}
\tag{16.23}
$$

$$
\begin{aligned}
792 \quad \text{off} \quad & 1+7\sqrt[13]{+1}+5\sqrt[13]{-1} \\
792 \quad \text{off} \quad & 1+7\sqrt[13]{-1}+5\sqrt[13]{+1}
\end{aligned}
\tag{16.24}
$$

$$
924 \quad \text{off} \quad 1+6\sqrt[13]{+1}+6\sqrt[13]{-1}
\tag{16.25}
$$

Being of prime order, there are no geometric spaces within the C_{13} group.

The order 14 cyclic group:

The order fourteen cyclic finite group, C_{14}, has two subgroups, $\{C_2, C_7\}$. The order fourteen cyclic group has 8,192 separate algebras. With the Standard form Cayley table used in this book, the form of these algebras is:

$$a + b\sqrt[14]{\pm 1} + c\sqrt[7]{\pm 1} + d\sqrt[14]{\pm 1} + e\sqrt[7]{\pm 1} + f\sqrt[14]{\pm 1} + g\sqrt[7]{\pm 1} + h\sqrt[2]{\pm 1} + i\sqrt[7]{\pm 1} + j\sqrt[14]{\pm 1} + k\sqrt[7]{\pm 1} + l\sqrt[14]{\pm 1} + m\sqrt[7]{\pm 1} + n\sqrt[14]{\pm 1} \quad (16.26)$$

There are fourteen different (non-isomorphic) division algebras. They are of the form:

$$
\begin{array}{lll}
64 & \text{off} & 1 + \sqrt[2]{+1} + 6\sqrt[7]{+1} + 6\sqrt[14]{+1} \\
64 & \text{off} & 1 + \sqrt[2]{-1} + 6\sqrt[7]{-1} + 6\sqrt[14]{-1}
\end{array}
\quad (16.27)
$$

$$
\begin{array}{lll}
64 & \text{off} & 1 + \sqrt[2]{+1} + 6\sqrt[7]{-1} + 6\sqrt[14]{+1} \\
64 & \text{off} & 1 + \sqrt[2]{-1} + 6\sqrt[7]{+1} + 6\sqrt[14]{-1}
\end{array}
\quad (16.28)
$$

$$
\begin{array}{lll}
384 & \text{off} & 1 + \sqrt[2]{+1} + 5\sqrt[7]{+1} + \sqrt[7]{-1} + 6\sqrt[14]{+1} \\
384 & \text{off} & 1 + \sqrt[2]{-1} + 5\sqrt[3]{-1} + \sqrt[7]{+1} + 6\sqrt[14]{-1}
\end{array}
\quad (16.29)
$$

$$
\begin{array}{lll}
384 & \text{off} & 1 + \sqrt[2]{+1} + 5\sqrt[7]{-1} + \sqrt[7]{+1} + 6\sqrt[14]{+1} \\
384 & \text{off} & 1 + \sqrt[2]{-1} + 5\sqrt[3]{+1} + \sqrt[7]{-1} + 6\sqrt[14]{-1}
\end{array}
\quad (16.30)
$$

$$
\begin{array}{lll}
960 & \text{off} & 1 + \sqrt[2]{-1} + 4\sqrt[7]{-1} + 2\sqrt[7]{+1} + 6\sqrt[14]{-1} \\
960 & \text{off} & 1 + \sqrt[2]{+1} + 4\sqrt[3]{+1} + 2\sqrt[7]{-1} + 6\sqrt[14]{+1}
\end{array}
\quad (16.31)
$$

$$
\begin{array}{lll}
960 & \text{off} & 1 + \sqrt[2]{-1} + 4\sqrt[7]{+1} + 2\sqrt[7]{-1} + 6\sqrt[14]{-1} \\
960 & \text{off} & 1 + \sqrt[2]{+1} + 4\sqrt[3]{-1} + 2\sqrt[7]{+1} + 6\sqrt[14]{+1}
\end{array}
\quad (16.32)
$$

$$
\begin{array}{lll}
1280 & \text{off} & 1 + \sqrt[2]{-1} + 3\sqrt[7]{+1} + 3\sqrt[7]{-1} + 6\sqrt[14]{-1} \\
1280 & \text{off} & 1 + \sqrt[2]{+1} + 4\sqrt[3]{-1} + 2\sqrt[7]{+1} + 6\sqrt[14]{+1}
\end{array}
\quad (16.33)
$$

Since the order fourteen cyclic group C_{14} contains a C_7 subgroup, we know that it cannot contain any geometric spaces.

The order 15 cyclic group:

The order fifteen cyclic finite group, C_{15}, has two subgroups, $\{C_3, C_5\}$. The order fifteen cyclic group has 16,384 separate algebras. With the Standard form Cayley table used in this book, the form of these algebras is:

$$a+b\sqrt[15]{\pm1}+c\sqrt[15]{\pm1}+d\sqrt[5]{\pm1}+e\sqrt[15]{\pm1}+f\sqrt[3]{\pm1}+g\sqrt[5]{\pm1}+h\sqrt[15]{\pm1}$$
$$+i\sqrt[15]{\pm1}+j\sqrt[5]{\pm1}+k\sqrt[3]{\pm1}+l\sqrt[15]{\pm1}+m\sqrt[7]{\pm1}+n\sqrt[15]{\pm1}+o\sqrt[15]{\pm1}$$

(16.34)

There are 125 different (non-isomorphic) division algebras within the C_{15} finite group. We do not list them.

Since the order fifteen cyclic group C_{15} contains a C_5 (or C_3) subgroup, we know that it cannot contain any geometric spaces.

Summary of the cyclic groups to order fifteen:

Your author has insufficient computer power available to him to check by brute force calculation any cyclic groups of order fifteen or above for emergent distance functions which form a geometric space.

By computer calculation, we have checked every cyclic group of order fifteen or less and we have found no geometric spaces within these lower order cyclic groups except for C_2. This means that no higher order group which has a cyclic group of order three through fifteen can contain any geometric spaces.

We have catalogued the division algebras which derive from the finite groups to order fifteen. The numbers and types are:

Group	Total No. of Algebras	No. of types	Distribution of different types of algebras
C_1	1	1	(1)
C_2	2	2	(1,1)
C_3	4	3	(1,2,1)
C_4	8	2	(4,4)
C_5	16	5	(1,4,6,4,1)
C_6	32	6	(4,4,8,8,4,4)
C_7	64	7	(1,6,15,20,15,6,1)
C_8	128	2	(64,64)
C_9	256	21	(1,1,2,6,6,12,15,15,20,30,40,30,20,15,15,12,6,6,2,1,1)
C_{10}	512	9	(16,16,64,64,96,64,64,16,16)
C_{11}	1,024	11	(1,10,45,120,210,252,210,120,45,10,1)

C_{12}	2,048	5	(256,256,512,256,256)
C_{13}	4,096	13	(1,12,66,220,495,792,924,792,495,220,66,12,1)
C_{14}	8,192	13	(64,64,384,384,960,960,1280,960,960,384,384,64,64)
C_{15}	16,384	125	Too many to present

Cyclic groups and geometric spaces:

We have tested the cyclic groups to order fifteen for emergent geometric spaces and found none. We know that all subgroups of cyclic groups are themselves cyclic. We know that, if the order, Ord, of a cyclic subgroup is divisible by an integer, d, then that cyclic group of order Ord has a cyclic subgroup of order d. Since any group that has a cyclic subgroup of order three through fifteen cannot hold a geometric space and no group of prime order other than C_2 can hold a geometric space, then almost every cyclic group cannot hold a geometric space.

The only cyclic groups that might hold a geometric space are those of order $2p$ where p is a prime number greater than fifteen; an example is C_{34}. However, we know that a group can hold a geometric space only if it is of order 2^n. Thus it is that no cyclic group of order greater than two holds a geometric space.

SECTION III – The Crossed Groups

Chapter 17

The Order Eight $C_2 \times C_2 \times C_2$ Group

General notes:

The group $C_m \times C_n$ is isomorphic to (the same group as) the group $C_n \times C_m$, and in general such permutation of subscripts always denote the same group.

If two groups, $\{G, H\}$ are abelian (commutative), then $G \times H$ is abelian.

The order of $G \times H$ is the product of the orders of $G \& H$.

If the highest common factor of the orders of two groups is unity, then the cross product of those two groups is isomorphic to the cyclic group of order equal to the product of the orders of the two groups. For example; $C_2 \times C_5 \cong C_{10}$.

There are two order eight groups formed by crossing cyclic groups together; these are the $C_2 \times C_2 \times C_2$ group and the $C_2 \times C_4$ group.

The $C_2 \times C_2 \times C_2$ group:

The $C_2 \times C_2 \times C_2$ Standard form Cayley table is:

$$
\begin{bmatrix}
a & b & c & d & e & f & g & h \\
b & a & d & c & f & e & h & g \\
c & d & a & b & g & h & e & f \\
d & c & b & a & h & g & f & e \\
e & f & g & h & a & b & c & d \\
f & e & h & g & b & a & d & c \\
g & h & e & f & c & d & a & b \\
h & g & f & e & d & c & b & a
\end{bmatrix}
\tag{17.1}
$$

We see that all the variables are symmetric across the leading diagonal. As a matrix, the Standard form Cayley table of the $C_2 \times C_2 \times C_2$ group has real eigenvalues and orthogonal eigenvectors.

The commutative $C_2 \times C_2 \times C_2$ group, like the commutative $C_2 \times C_2$ group, has both commutative and non-commutative division algebras within it. The non-commutative algebras arise from three quadratic equations necessary to eliminate sufficient parameters to give a multiplicatively closed algebraic matrix form analogously to the $C_2 \times C_2$ group presented above.

The non-commutative $C_2 \times C_2 \times C_2$ division algebras (spinor algebras) are the division algebra forms of the 8-dimensional Clifford algebras.

The $C_2 \times C_2 \times C_2$ group has seven C_2 subgroups and seven $C_2 \times C_2$ subgroups. The seven C_2 subgroups are the identity, a, and each of the other variables. The seven $C_2 \times C_2$ subgroups are: $\{a,b,c,d\}$, $\{a,b,e,f\}$, $\{a,b,g,h\}$, $\{a,c,e,g\}$, $\{a,c,f,h\}$, $\{a,d,e,h\}$ and $\{a,d,f,g\}$.

The $C_2 \times C_2 \times C_2$ commutative algebraic matrix form:

The commutative $C_2 \times C_2 \times C_2$ algebraic matrix form is:

$$\textit{Commutative}$$

$$
\begin{bmatrix}
a & b & c & d & e & f & g & h \\[4pt]
P_{2.1}b & a & P_{2.3}d & \dfrac{P_{2.1}}{P_{2.3}}c & P_{2.5}f & \dfrac{P_{2.1}}{P_{2.5}}e & P_{2.7}h & \dfrac{P_{2.1}}{P_{2.7}}g \\[10pt]
P_{3.1}c & \dfrac{P_{2.3}P_{3.1}}{P_{2.1}}d & a & \dfrac{P_{2.1}}{P_{2.3}}b & P_{3.5}g & \dfrac{P_{2.7}P_{3.5}}{P_{2.5}}h & \dfrac{P_{3.1}}{P_{3.5}}e & \dfrac{P_{2.5}P_{3.1}}{P_{2.7}P_{3.5}}f \\[10pt]
\dfrac{P_{2.3}{}^2 P_{3.1}}{P_{2.1}}d & \dfrac{P_{2.3}P_{3.1}}{P_{2.1}}c & P_{2.3}b & a & \dfrac{P_{2.3}P_{2.7}P_{3.5}}{P_{2.1}}h & \dfrac{P_{2.3}P_{3.5}}{P_{2.5}}g & \dfrac{P_{2.3}P_{2.5}P_{3.1}}{P_{2.1}P_{3.5}}f & \dfrac{P_{2.3}P_{3.1}}{P_{2.7}P_{3.5}}e \\[10pt]
P_{5.1}e & \dfrac{P_{2.5}P_{5.1}}{P_{2.1}}f & \dfrac{P_{3.5}P_{5.1}}{P_{3.1}}g & \dfrac{P_{2.7}P_{3.5}P_{5.1}}{P_{2.3}P_{3.1}}h & a & \dfrac{P_{2.1}}{P_{2.5}}b & \dfrac{P_{3.1}}{P_{3.5}}c & \dfrac{P_{2.3}P_{3.1}}{P_{2.7}P_{3.5}}d \\[10pt]
\dfrac{P_{2.5}{}^2 P_{5.1}}{P_{2.1}}f & \dfrac{P_{2.5}P_{5.1}}{P_{2.1}}e & \dfrac{P_{2.5}P_{2.7}P_{3.5}P_{5.1}}{P_{2.1}P_{3.1}}h & \dfrac{P_{2.5}P_{3.5}P_{5.1}}{P_{2.3}P_{3.1}}g & P_{2.5}b & a & \dfrac{P_{2.3}P_{2.5}P_{3.1}}{P_{2.1}P_{3.5}}d & \dfrac{P_{2.5}P_{3.1}}{P_{2.7}P_{3.5}}c \\[10pt]
\dfrac{P_{3.5}{}^2 P_{5.1}}{P_{3.1}}g & \dfrac{P_{2.7}P_{3.5}{}^2 P_{5.1}}{P_{2.1}P_{3.1}}h & \dfrac{P_{3.5}P_{5.1}}{P_{3.1}}e & \dfrac{P_{2.5}P_{3.5}P_{5.1}}{P_{2.3}P_{3.1}}f & P_{3.5}c & \dfrac{P_{2.3}P_{3.5}}{P_{2.5}}d & a & \dfrac{P_{2.1}}{P_{2.7}}b \\[10pt]
\dfrac{P_{2.7}{}^2 P_{3.5}{}^2 P_{5.1}}{P_{2.1}P_{3.1}}h & \dfrac{P_{2.7}P_{3.5}{}^2 P_{5.1}}{P_{2.1}P_{3.1}}g & \dfrac{P_{2.5}P_{2.7}P_{3.5}P_{5.1}}{P_{2.1}P_{3.1}}f & \dfrac{P_{2.7}P_{3.5}P_{5.1}}{P_{2.3}P_{3.1}}e & \dfrac{P_{2.3}P_{2.7}P_{3.5}}{P_{2.1}}d & \dfrac{P_{2.7}P_{3.5}}{P_{2.5}}c & P_{2.7}b & a
\end{bmatrix}
$$

$$(17.2)$$

We see there are the seven scaling parameters, $P_{2.1}$, $P_{2.3}$, $P_{2.5}$, $P_{2.7}$, $P_{3.1}$, $P_{3.5}$, $P_{5.1}$. Setting each of these seven scaling parameters to either ± 1 gives $2^7 = 128$ different permutations which are 128 division algebras. Many of these algebras are algebraically isomorphic.

The $C_2 \times C_2 \times C_2$ non-commutative algebraic matrix form:

Over and above these 128 permutations of the scaling parameters, there are the various distributions of minus signs due to extracting the negative roots of the three quadratic equations that arise in the calculation that eliminates the unnecessary scaling parameters. There are eight such distributions given by the eight permutations of setting the three S_i to ± 1. Those distributions are:

$$
\begin{bmatrix}
a & b & c & d & e & f & g & h \\
\sim & a & \sim & \sim & \sim & \sim & \sim & \sim \\
\sim & S_4 & a & S_4 & \sim & S_4 & \sim & S_4 \\
S_4 & \sim & S_4 & a & S_4 & \sim & S_4 & \sim \\
\sim & S_6 S_7 & S_7 & S_6 & a & S_6 S_7 & S_7 & S_6 \\
S_6 S_7 & \sim & S_6 & S_7 & S_6 S_7 & a & S_6 & S_7 \\
S_7 & S_4 S_6 & \sim & S_4 S_6 S_7 & S_7 & S_4 S_6 & a & S_4 S_6 S_7 \\
S_4 S_6 & S_7 & S_4 S_6 S_7 & \sim & S_4 S_6 & S_7 & S_4 S_6 S_7 & a
\end{bmatrix}
\tag{17.3}
$$

For example, with $P_{2.1} = P_{2.3} = P_{2.5} = P_{2.7} = P_{3.1} = P_{3.5} = P_{5.1} = 1$, the permutation $S_4 = -1$, $S_6 = +1$, $S_7 = +1$ gives the distribution of minus signs:

$$
\exp\left(
\begin{bmatrix}
a & b & c & d & e & f & g & h \\
b & a & d & c & f & e & h & g \\
c & -d & a & -b & g & -h & e & -f \\
-d & c & -b & a & -h & g & -f & e \\
e & f & g & h & a & b & c & d \\
f & e & h & g & b & a & d & c \\
g & -h & e & -f & c & -d & a & -b \\
-h & g & -f & e & -d & c & -b & a
\end{bmatrix}
\right)
\tag{17.4}
$$

Thus, there are 128 algebras for each of the $2^3 = 8$ permutations of the $S_i = \pm 1$, and so we have in total $8 \times 128 = 1024$ 8-dimensional $C_2 \times C_2 \times C_2$ division algebras. However, the seven permutations of $S_i = \pm 1$ other than $S_i = +1$ give the same sets of algebras.

Note: Part of the computer code used to form the $C_2 \times C_2 \times C_2$ algebraic matrix form is:

> $P[4, 1] := \dfrac{(-1)^{PM4} \cdot P[2, 3]^2 \cdot P[3, 1]}{P[2, 1]}$: $P[6, 1] := \dfrac{(-1)^{PM6} \cdot P[2, 5]^2 \cdot P[3, 1] \cdot P[7, 1]}{P[2, 1] \cdot P[3, 5]^2}$:

> $P[7, 1] := \dfrac{(-1)^{PM7} \cdot P[3, 5]^2 \cdot P[5, 1]}{P[3, 1]}$:

The signs S_i are given the subscript corresponding to the parameter they are associated with in the above code; for example, S_4 appears in the code is associated with the elimination of the $P_{4,1}$ parameter.

So, to summarise: There are 128 commutative division algebras and seven equivalent sets of 128 non-commutative algebras within the $C_2 \times C_2 \times C_2$ group.

The eight algebraic matrix forms of $C_2 \times C_2 \times C_2$:

Looking at (17.3), we see that we have eight different algebraic matrix forms of the $C_2 \times C_2 \times C_2$ group. However, since the seven non-commutative forms produce exactly the same sets of algebras, we present only one of the seven non-commutative algebraic matrix forms of the $C_2 \times C_2 \times C_2$ group. A non-commutative algebraic matrix form of the $C_2 \times C_2 \times C_2$ group is:

$$Non-Commutative$$

$$
\begin{bmatrix}
a & b & c & d & e & f & g & h \\[4pt]
P_{2,1}b & a & P_{2,3}d & \dfrac{P_{2,1}}{P_{2,3}}c & P_{2,5}f & \dfrac{P_{2,1}}{P_{2,5}}e & P_{2,7}h & \dfrac{P_{2,1}}{P_{2,7}}g \\[10pt]
P_{3,1}c & -\dfrac{P_{2,3}P_{3,1}}{P_{2,1}}d & a & -\dfrac{P_{2,1}}{P_{2,3}}b & P_{3,5}g & -\dfrac{P_{2,7}P_{3,5}}{P_{2,5}}h & \dfrac{P_{3,1}}{P_{3,5}}e & -\dfrac{P_{2,5}P_{3,1}}{P_{2,7}P_{3,5}}f \\[12pt]
-\dfrac{P_{2,3}^2 P_{3,1}}{P_{2,1}}d & \dfrac{P_{2,3}P_{3,1}}{P_{2,1}}c & -P_{2,3}b & a & -\dfrac{P_{2,3}P_{2,7}P_{3,5}}{P_{2,1}}h & \dfrac{P_{2,3}P_{3,5}}{P_{2,5}}g & -\dfrac{P_{2,3}P_{2,5}P_{3,1}}{P_{2,1}P_{3,5}}f & \dfrac{P_{2,3}P_{3,1}}{P_{2,7}P_{3,5}}e \\[12pt]
P_{5,1}e & \dfrac{P_{2,5}P_{5,1}}{P_{2,1}}f & \dfrac{P_{3,5}P_{5,1}}{P_{3,1}}g & \dfrac{P_{2,7}P_{3,5}P_{5,1}}{P_{2,3}P_{3,1}}h & a & \dfrac{P_{2,1}}{P_{2,5}}b & \dfrac{P_{3,1}}{P_{3,5}}c & \dfrac{P_{2,3}P_{3,1}}{P_{2,7}P_{3,5}}d \\[12pt]
\dfrac{P_{2,5}^2 P_{5,1}}{P_{2,1}}f & \dfrac{P_{2,5}P_{5,1}}{P_{2,1}}e & \dfrac{P_{2,5}P_{2,7}P_{3,5}P_{5,1}}{P_{2,1}P_{3,1}}h & \dfrac{P_{2,5}P_{3,5}P_{5,1}}{P_{2,3}P_{3,1}}g & P_{2,5}b & a & \dfrac{P_{2,3}P_{2,5}P_{3,1}}{P_{2,1}P_{3,5}}d & \dfrac{P_{2,5}P_{3,1}}{P_{2,7}P_{3,5}}c \\[12pt]
\dfrac{P_{3,5}^2 P_{5,1}}{P_{3,1}}g & -\dfrac{P_{2,7}P_{3,5}^2 P_{5,1}}{P_{2,1}P_{3,1}}h & \dfrac{P_{3,5}P_{5,1}}{P_{3,1}}e & -\dfrac{P_{2,5}P_{3,5}P_{5,1}}{P_{2,3}P_{3,1}}f & P_{3,5}c & -\dfrac{P_{2,3}P_{3,5}}{P_{2,5}}d & a & -\dfrac{P_{2,1}}{P_{2,7}}b \\[12pt]
-\dfrac{P_{2,7}P_{3,5}^2 P_{5,1}}{P_{2,1}P_{3,1}}h & \dfrac{P_{2,7}P_{3,5}^2 P_{5,1}}{P_{2,1}P_{3,1}}g & -\dfrac{P_{2,5}P_{2,7}P_{3,5}P_{5,1}}{P_{2,1}P_{3,1}}f & \dfrac{P_{2,7}P_{3,5}P_{5,1}}{P_{2,3}P_{3,1}}e & -\dfrac{P_{2,3}P_{2,7}P_{3,5}}{P_{2,1}}d & \dfrac{P_{2,7}P_{3,5}}{P_{2,5}}c & -P_{2,7}b & a
\end{bmatrix}
$$

$$(17.5)$$

4-dimensional sub-algebras:

Let us compare the top left-hand 4×4 block of (17.5) with the algebraic matrix form of the $C_2 \times C_2$ given earlier (7.3) reproduced here for the reader's convenience:

$$
C_2 \times C_2^{\,Non\text{-}Commutative} \sim
\begin{bmatrix}
a & b & c & d \\[6pt]
P_{2,1}b & a & P_{2,3}d & \dfrac{P_{2,1}}{P_{2,3}}c \\[10pt]
P_{3,1}c & -\dfrac{P_{2,3}P_{3,1}}{P_{2,1}}d & a & -\dfrac{P_{2,1}}{P_{2,3}}b \\[12pt]
-\dfrac{(P_{2,3})^2 P_{3,1}}{P_{2,1}}d & \dfrac{P_{2,3}P_{3,1}}{P_{2,1}}c & -P_{2,3}b & a
\end{bmatrix}
\qquad (17.6)
$$

We see that we have an exact match. Putting $P_{2,1} = P_{2,3} = -1$, $P_{3,1} = +1$ into the 4-dimensional algebraic matrix form, (17.6), gives an A_3 algebra:

$$ASS \sim \begin{bmatrix} a & b & c & d \\ -b & a & -d & c \\ c & -d & a & -b \\ d & c & b & a \end{bmatrix}$$

(17.7)

$$\det(ASS) = \left(a^2 + b^2 - c^2 - d^2\right)^2$$

To achieve the same in the top left-hand 4×4 block of the 8-dimensional algebraic matrix form, we put $P_{2,1} = P_{2,3} = -1$, $P_{3,1} = +1$. To achieve similar in the bottom right-hand corner, we put $P_{2,5} = P_{2,7} = P_{3,5} = -1$; this gives:

$$\begin{bmatrix} a & b & c & d & e & f & g & h \\ -b & a & -d & c & -f & e & -h & g \\ c & -d & a & -b & -g & h & -e & f \\ d & c & b & a & -h & -g & -f & -e \\ P_{5.1}e & P_{5.1}f & -P_{5.1}g & -P_{5.1}h & a & b & -c & -d \\ -P_{5.1}f & P_{5.1}e & P_{5.1}h & -P_{5.1}g & -b & a & d & -c \\ P_{5.1}g & -P_{5.1}h & -P_{5.1}e & P_{5.1}f & -c & d & a & -b \\ P_{5.1}h & P_{5.1}g & -P_{5.1}f & -P_{5.1}e & -d & -c & b & a \end{bmatrix}$$

(17.8)

Setting the $e = f = g = h = 0$ variables to zero in (17.8) produces a matrix whose determinant is:

$$dist^8 = \left(a^2 + b^2 - c^2 - d^2\right)^4$$

(17.9)

In fact, with any combination of signs ± 1 for the three parameters $P_{2,5} = P_{2,7} = P_{3,5} = \pm 1$, we get the above determinant. Further, the eight different permutations of the $P_{2,1}, P_{2,3}, P_{3,1} = \pm 1$ give the six A_3 algebras and the two quaternion algebras. Clearly, we might have the potential for 4-dimensional emergent geometric spaces when four of the eight variables are zero. However, things are not that simple.

Other 4-dimensional sub-algebras:

Because the $C_2 \times C_2 \times C_2$ group has seven $C_2 \times C_2$ subgroups, the 8-dimensional algebras will become 4-dimensional algebras when seven sets of four non-identity variables are set to zero. Of these seven sub-algebras, only four, the $\{a,b,c,d\}$, $\{a,b,g,h\}$, $\{a,c,f,h\}$, and $\{a,d,f,g\}$ have determinant of the form (17.9) given above. The other three of the seven sub-algebras have determinants of the form:

$$dist^8 = \left(\left(a-e\right)^2 + \left(b+f\right)^2\right)^2 \left(\left(a+e\right)^2 + \left(b-f\right)^2\right)^2$$

(17.10)

Looking back at the commutative $C_2 \times C_2$ algebras, circa (7.7), we see that these three sub-algebras are commutative sub-algebras.

The $C_2 \times C_2 \times C_2$ algebras:

The signs permutation $S_4 = S_6 = S_7 = +1$ give 128 commutative algebras. Of these, there are 112 algebras of the form $1 + 3\sqrt{+1} + 4\sqrt{-1}_{Com}$ and 16 algebras of the form $1 + 7\sqrt{+1}$. The other seven permutations of S_i each give one of seven sets of 128 non-commutative algebras. The algebras in each set match as far as algebraic isomorphism. Each set is comprised of 16 algebras of the form $1 + 1\sqrt{+1} + 6\sqrt{-1}$ with 48 algebras of the form $1 + 5\sqrt{+1} + 2\sqrt{-1}$ and 64 algebras of the form $1 + 3\sqrt{+1} + 4\sqrt{-1}_{Non-Comm}$. Of course, all the algebras of a particular type are algebraically isomorphic but are written in different bases. Notice we have both commutative and non-commutative versions of the $1 + 3\sqrt{+1} + 4\sqrt{-1}$ algebras.

The $1 + 7\sqrt{+1}$ algebras:

There are sixteen $1 + 7\sqrt{+1}$ algebras; these are commutative algebras.

There are no geometric spaces emerge from these algebras.

The commutative $1 + 3\sqrt{+1} + 4\sqrt{-1}_{Com}$ algebras:

There are one hundred and twelve $1 + 3\sqrt{+1} + 4\sqrt{-1}_{Com}$ algebras; these are commutative algebras.

There are no geometric spaces emerge from these algebras.

The non-commutative $1 + 3\sqrt{+1} + 4\sqrt{-1}_{Non-Com}$ algebras:

There are seven sets of 128 non-commutative algebras corresponding to the seven permutations of S_i which include at least one minus unity. There are sixty-four $1 + 3\sqrt{+1} + 4\sqrt{-1}_{Non-Com}$ algebras in each set.

There are no geometric spaces emerge from these algebras.

The non-commutative $1 + 5\sqrt{+1} + 2\sqrt{-1}$ algebras:

There are seven sets of 48 non-commutative $1 + 5\sqrt{+1} + 2\sqrt{-1}$ algebras corresponding to the seven permutations of S_i which include at least one minus unity.

There are no geometric spaces emerge from these algebras.

The non-commutative $1 + 1\sqrt{+1} + 6\sqrt{-1}$ algebras:

There are seven sets of 16 non-commutative $1 + 1\sqrt{+1} + 6\sqrt{-1}$ algebras corresponding to the seven permutations of S_i which include at least one minus unity.

By the standard definition, there are no geometric spaces emerge from these algebras.

However, there is a complication of which we feel we should make the reader aware. This complication might be connected to quarks and the strong force although, at this stage, this is no more than wild speculation.

In search of the strong force perhaps. More geometric spaces perhaps:

The algebras of the $C_2 \times C_2 \times C_2$ group have the interesting property of 'tying together' the eight variables into four pairs. We see this in both the determinants (see below) and the trigonometric functions of these algebras. This 'tying together' of variables happens because the sub-group structure of these algebras, seven 2-dimensional sub-algebras and seven 4-dimensional sub-algebras, needs to be satisfied. The $C_2 \times C_2 \times C_2 \times C_2$ algebras also 'tie together' pairs of pairs of variables. The $C_2 \times C_2 \times C_2 \times C_2 \times C_2$ algebras also 'tie together' pairs of pairs of pairs of variables. This is a general property of this type of group.

The $1 + 6\sqrt[2]{-1} + \sqrt[2]{+1}$ algebra:

The $1 + 6\sqrt[2]{-1} + \sqrt[2]{+1}$ algebras have determinants of the form:

$$\det = \left((a+e)^2 + (b-f)^2 + (c+g)^2 + (d-h)^2 \right)^2 \left((a-e)^2 + (b+f)^2 + (c-g)^2 + (d+h)^2 \right)^2$$

(17.11)

We see the pairing of variables. This is equal to the 8-dimensional distance function. We can take the square root of this to give a 4-dimensional distance function.

$$dist^4 = \left((a+e)^2 + (b-f)^2 + (c+g)^2 + (d-h)^2 \right)\left((a-e)^2 + (b+f)^2 + (c-g)^2 + (d+h)^2 \right)$$

(17.12)

Multiplying this out gives:

$$\begin{aligned}
dist^4 &= \left(a^2 - e^2\right)^2 + (ab+af+be+ef)^2 + (ac-ag+ce-eg)^2 + (ad+ah+de+eh)^2 \\
&\quad (ab-be-af+ef)^2 + \left(b^2-f^2\right)^2 + (bc-bg-cf+fg)^2 + (bd+bh-df-fh)^2 \\
&\quad (ac-ce+ag-eg)^2 + (bc+cf+bg+fg)^2 + \left(c^2-g^2\right)^2 + (cd+ch+dg+gh)^2 \\
&\quad (ad-ed-ah+eh)^2 + (bd+df-bh-fh)^2 + (cd-dg-ch+gh)^2 + \left(d^2-h^2\right)^2
\end{aligned}$$

(17.13)

If we allow the tangling together of the different variables, this is of the form of a 16-dimensional Euclidean distance function (a quadratic form with sixteen squared quantities) but it equates to a 4th power of distance. Altogether, this tangling together of variables is unacceptable to our minds, and so we reject it.

If we put $a = e$, $b = f$, $c = g$, $d = h$, we get:

$$dist^4 = (4ab)^2 + (4ad)^2 + (4bc)^2 + (4cd)^2$$

(17.14)

This is closer, but this is disqualified from being a geometric space by the distance being a 4th power.

However, there are two other non-commutative algebras within the $C_2 \times C_2 \times C_2$ group. Perhaps, with the idea of having variables tied together as they appear to be in the determinants, like (17.11), and in the trigonometric functions of these algebras, we should look at those algebras. We will assume $a = e$, $b = f$, $c = g$, $d = h$.

We begin by saying that, even with the variables 'tied together' as $a = e$, $b = f$, $c = g$, $d = h$, the $1 + 2\sqrt[2]{-1} + 5\sqrt[2]{+1}$ will not form a geometric space.

However, the $1 + 4\sqrt[2]{-1} + 3\sqrt[2]{+1}$ algebras, if the variables are 'tied together' as $a = e$, $b = f$, $c = g$, $d = h$ do form prospective geometric spaces.

The $1 + 4\sqrt[2]{-1} + 3\sqrt[2]{+1}$ algebras:

With $a = e$, $b = f$, $c = g$, $d = h$, the $1 + 4\sqrt[2]{-1} + 3\sqrt[2]{+1}$ algebras have determinants of the forms:

$$dist^8 = \left(a^2 + b^2 + c^2 + d^2\right)^4 \qquad 8 \quad off \tag{17.15}$$

$$dist^8 = \left(a^2 - b^2 + c^2 + d^2\right)^4 \qquad 8 \quad off$$

$$dist^8 = \left(a^2 + b^2 - c^2 + d^2\right)^4 \qquad 8 \quad off \tag{17.16}$$

$$dist^8 = \left(a^2 + b^2 + c^2 - d^2\right)^4 \qquad 8 \quad off$$

$$dist^8 = \left(a^2 - b^2 - c^2 + d^2\right)^4 \qquad 8 \quad off$$

$$dist^8 = \left(a^2 - b^2 + c^2 - d^2\right)^4 \qquad 8 \quad off \tag{17.17}$$

$$dist^8 = \left(a^2 + b^2 - c^2 - d^2\right)^4 \qquad 8 \quad off$$

$$dist^8 = \left(a^2 - b^2 - c^2 - d^2\right)^4 \qquad 8 \quad off \tag{17.18}$$

We can clearly take the 4th roots of these to give a quadratic distance function. After doing that, we can form the expectation distance function by adding determinants of the same type.

Adding algebras of the type (17.15) gives:

$$dist^2 = a^2 + b^2 + c^2 + d^2 \tag{17.19}$$

This is a quaternion type of geometric space.

Adding the algebras of the type (17.16) gives:

$$dist^2 = 3a^2 + b^2 + c^2 + d^2 \tag{17.20}$$

Because this is a Euclidean type of distance function without time, we cannot adjust the units to be rid of the 3. This is not a geometric space.

Adding the algebras of the type (17.17) gives:

$$dist^2 = 3a^2 - b^2 - c^2 - d^2 \qquad (17.21)$$

Because this is a distance function with time, we can adjust the units to be rid of the 3. This is the geometric space that is our space-time.

Adding the algebras of the type (17.18) gives:

$$dist^2 = a^2 - b^2 - c^2 - d^2 \qquad (17.22)$$

This again is our space-time.

Thus we see that, if we are prepared to pair the variables by putting $a=e$, $b=f$, $c=g$, $d=h$, we do find geometric spaces within the order eight $C_2 \times C_2 \times C_2$ group. Clearly, if the variables were to vary away from the $a=e$, $b=f$, $c=g$, $d=h$ condition, the 4-dimensional geometric spaces would be 'distorted' in some way. We wonder if this is connected to the strong force and quark confinement. Within quantum field theory, we associate electric charge with the imaginary variable of the complex numbers, $\mathbb{C}$. Suppose we associate the $\{a,b,c,d\}$ variables with electric charge of $\frac{2}{3}$ and we associate the $\{e,f,g,h\}$ variables with electric charge of $\frac{1}{3}$, then, when $a=e$, $b=f$, $c=g$, $d=h$ the $(a-e)$ type of terms vanish and the $(a+e)$ type of terms have electric charge unity. What are two differently charged imaginary variables in 8-dimensions become one single charge in 4-dimensions.

Is all this connected to the strong force? We do not state an opinion; we merely make the reader aware of that these 'folded' 8-dimensional spaces can form geometric spaces. Of course, the geometric spaces these 'folded' algebras form match the geometric spaces of the 4-dimension $C_2 \times C_2$ group – nothing new here.

The Computer code used to form the $C_2 \times C_2 \times C_2$ group algebras:

The code used to form the $C_2 \times C_2 \times C_2$ algebraic matrix form is:

```
>  R1 := [a, b, c, d, e, f, g, h] :
>  R2 := [b, a, d, c, f, e, h, g] :
>  R3 := [c, d, a, b, g, h, e, f] :
>  R4 := [d, c, b, a, h, g, f, e] :
>  R5 := [e, f, g, h, a, b, c, d] :
>  R6 := [f, e, h, g, b, a, d, c] :
>  R7 := [g, h, e, f, c, d, a, b] :
>  R8 := [h, g, f, e, d, c, b, a] :
>
>  a := A[0] :  b := A[1] :  c := A[2] :  d := A[3] : e := A[4] : f := A[5] :
>  g := A[6] : h := A[7] :
>  AN := Matrix([R1, R2, R3, R4, R5, R6, R7, R8]) :
>  VARMAT := copy(AN) :
>  a := S[0] : b := S[1] :  c := S[2] :  d := S[3] :  e := S[4] :  f := S[5] :
>  g := S[6] : h := S[7] :
```

```
> SY := Matrix([R1, R2, R3, R4, R5, R6, R7, R8]) :
> a := X[0] : b := X[1] : c := X[2] : d := X[3] : e := X[4] : f := X[5] :
> g := X[6] : h := X[7] :
> VARMAT2 := Matrix([R1, R2, R3, R4, R5, R6, R7, R8]) :
> a := B[0] : b := B[1] : c := B[2] : d := B[3] : e := B[4] : f := B[5] :
> g := B[6] : h := B[7] :
> VARMAT3 := Matrix([R1, R2, R3, R4, R5, R6, R7, R8]) :
> a := C[0] : b := C[1] : c := C[2] : d := C[3] : e := C[4] : f := C[5] :
> g := C[6] : h := C[7] :
> VARMAT4 := Matrix([R1, R2, R3, R4, R5, R6, R7, R8]) :
>
> for row from 1 to SIZE do
> for col from 1 to SIZE do
> ANDY[row, col] := P[row, col]·AN[row, col]
> SYD[row, col] := P[row, col]·SY[row, col]
> PARAM[row, col] := P[row, col] :
> end do: end do:
> for xx from 1 to SIZE do
> P[1, xx] := 1 :
> P[xx, xx] := 1 :
> end do:
> VARMAT;
```

$$> \quad P[4,3] := \frac{P[2,1]}{P[3,4]} \ : \quad P[6,5] := \frac{P[2,1]}{P[5,6]} \ : \ P[8,7] := \frac{P[2,1]}{P[7,8]} \ :$$

$$> \quad P[4,2] := \frac{P[3,1]}{P[2,4]} \ : \quad P[7,5] := \frac{P[3,1]}{P[5,7]} \ : \ P[8,6] := \frac{P[3,1]}{P[6,8]} \ :$$

$$> \quad P[3,2] := \frac{P[4,1]}{P[2,3]} \ : \ P[8,5] := \frac{P[4,1]}{P[5,8]} \ : \ P[7,6] := \frac{P[4,1]}{P[6,7]} \ :$$

$$> \quad P[6,2] := \frac{P[5,1]}{P[2,6]} \ : \ P[7,3] := \frac{P[5,1]}{P[3,7]} \ : \ P[8,4] := \frac{P[5,1]}{P[4,8]} \ :$$

$$> \quad P[5,2] := \frac{P[6,1]}{P[2,5]} \ : \ P[8,3] := \frac{P[6,1]}{P[3,8]} \ : \ P[7,4] := \frac{P[6,1]}{P[4,7]} \ :$$

$$> \quad P[8,2] := \frac{P[7,1]}{P[2,8]} \ : \ P[5,3] := \frac{P[7,1]}{P[3,5]} \ : \ P[6,4] := \frac{P[7,1]}{P[4,6]} \ :$$

$$> \quad P[7,2] := \frac{P[8,1]}{P[2,7]} \ : \ P[6,3] := \frac{P[8,1]}{P[3,6]} \ : \ P[5,4] := \frac{P[8,1]}{P[4,5]} \ :$$

```
>
```

$$> \quad P[3,4] := \frac{P[2,3]\cdot P[3,1]}{P[4,1]} \ : \quad P[2,4] := \frac{P[2,1]}{P[2,3]} \ : \quad P[6,8] := \frac{P[3,1]\cdot P[6,1]}{P[3,8]\,P[8,1]} \ :$$

$$> \quad P[5,7] := \frac{P[3,8]\cdot P[4,7]\cdot P[8,1]}{P[4,5]\cdot P[6,1]} \ : \quad P[4,7] := \frac{P[2,3]\cdot P[2,7]\cdot P[3,8]}{P[2,1]} \ :$$

$$> \quad P[3,8] := \frac{P[3,1]}{P[3,6]} \ : \quad P[4,5] := \frac{P[2,3]\cdot P[2,5]\cdot P[3,6]}{P[2,1]} \ : P[3,7] := \frac{P[3,1]}{P[3,5]} \ :$$

$$> \quad P[5,8] := \frac{P[4,1]\cdot P[5,1]}{P[4,8]\cdot P[8,1]} \ : \quad P[4,8] := \frac{P[2,1]\cdot P[4,1]}{P[2,3]\cdot P[2,5]\cdot P[3,6]} \ : \ P[2,6] := \frac{P[2,1]}{P[2,5]} \ :$$

$$> \quad P[7,8] := \frac{P[2,1]\cdot P[7,1]}{P[2,8]\cdot P[8,1]} \ : P[6,7] := \frac{P[4,6]\cdot P[6,1]}{P[7,1]} \ : \ P[5,6] := \frac{P[2,5]\cdot P[5,1]}{P[6,1]} \ :$$

$$> \quad P[2,8] := \frac{P[2,1]}{P[2,7]} \ : P[4,6] := \frac{P[2,3]\cdot P[3,5]}{P[2,5]} \ :$$

$$> \quad P[3,6] := \frac{P[2,1]\cdot P[2,7]\cdot P[3,5]\cdot P[4,1]}{P[2,3]^2\cdot P[2,5]\cdot P[3,1]}$$

> $P[8,1] := \dfrac{P[2,5]^2 \cdot P[2,7]^2 \cdot P[4,1] \cdot P[5,1] \cdot P[7,1]}{P[2,1] \cdot P[2,3]^2 \cdot P[3,1] \cdot P[6,1]}$:

>

> $PM4 := 1$:   *# Setting PM4, PM6, & PM7 to 1 or zero produces the eight sets of algebras.*
> $PM6 := 0$:
> $PM7 := 0$:

>

> $P[4,1] := \dfrac{(-1)^{PM4} \cdot P[2,3]^2 \cdot P[3,1]}{P[2,1]}$: $P[6,1] := \dfrac{(-1)^{PM6} \cdot P[2,5]^2 \cdot P[3,1] \cdot P[7,1]}{P[2,1] \cdot P[3,5]^2}$:

> $P[7,1] := \dfrac{(-1)^{PM7} \cdot P[3,5]^2 \cdot P[5,1]}{P[3,1]}$:

The Order Eight $C_2 \times C_4$ Group

The $C_2 \times C_4$ group is a commutative group which has three C_2 subgroup and two C_4 subgroups. Its Standard form Cayley table is of the form of four copies of the cyclic group C_4:

$$C_2 \times C_4 = \begin{bmatrix} a & b & c & d & e & f & g & h \\ d & a & b & c & h & e & f & g \\ c & d & a & b & g & h & e & f \\ b & c & d & a & f & g & h & e \\ e & f & g & h & a & b & c & d \\ h & e & f & g & d & a & b & c \\ g & h & e & f & c & d & a & b \\ f & g & h & e & b & c & d & a \end{bmatrix} \tag{18.1}$$

The $C_2 \times C_4$ group has algebraic matrix form:

$$C_2 \times C_4 =$$

$$\begin{bmatrix} a & b & c & d & e & f & g & h \\[2ex] P_{2,1}d & a & P_{2,3}b & \dfrac{P_{2,3}P_{3,1}}{P_{2,1}}c & P_{2,5}h & P_{2,6}e & P_{2,7}f & \dfrac{P_{2,3}{}^2 P_{3,1}}{P_{2,5}P_{2,6}P_{2,7}}g \\[3ex] P_{3,1}c & \dfrac{P_{2,1}}{P_{2,3}}d & a & \dfrac{P_{2,3}P_{3,1}}{P_{2,1}}b & \dfrac{P_{2,3}P_{3,1}}{P_{2,6}P_{2,7}}g & \dfrac{P_{2,5}P_{2,6}}{P_{2,3}}h & \dfrac{P_{2,6}P_{2,7}}{P_{2,3}}e & \dfrac{P_{2,3}P_{3,1}}{P_{2,5}P_{2,6}}f \\[3ex] P_{2,1}b & \dfrac{P_{2,1}}{P_{2,3}}c & \dfrac{(P_{2,1})^2}{P_{2,3}P_{3,1}}d & a & \dfrac{P_{2,1}}{P_{2,6}}f & \dfrac{P_{2,1}}{P_{2,7}}g & \dfrac{P_{2,1}P_{2,5}P_{2,6}P_{2,7}}{P_{2,3}{}^2 P_{3,1}}h & \dfrac{P_{2,1}}{P_{2,5}}e \\[3ex] P_{5,1}e & \dfrac{P_{5,1}}{P_{2,6}}f & \dfrac{P_{2,3}P_{5,1}}{P_{2,6}P_{2,7}}g & \dfrac{P_{2,5}P_{5,1}}{P_{2,1}}h & a & P_{2,6}b & \dfrac{P_{2,6}P_{2,7}}{P_{2,3}}c & \dfrac{P_{2,1}}{P_{2,5}}d \\[3ex] \dfrac{P_{2,5}P_{5,1}}{P_{2,6}}h & \dfrac{P_{5,1}}{P_{2,6}}e & \dfrac{P_{2,3}P_{5,1}}{P_{2,6}{}^2}f & \dfrac{P_{2,3}{}^2 P_{3,1}P_{5,1}}{P_{2,1}P_{2,6}{}^2 P_{2,7}}g & \dfrac{P_{2,1}}{P_{2,6}}d & a & P_{2,7}b & \dfrac{P_{2,3}P_{3,1}}{P_{2,5}P_{2,6}}c \\[3ex] \dfrac{P_{2,3}{}^2 P_{3,1}P_{5,1}}{P_{2,6}{}^2 P_{2,7}{}^2}g & \dfrac{P_{2,5}P_{5,1}}{P_{2,6}P_{2,7}}h & \dfrac{P_{2,3}P_{5,1}}{P_{2,6}P_{2,7}}e & \dfrac{P_{2,3}{}^2 P_{3,1}P_{5,1}}{P_{2,1}P_{2,6}{}^2 P_{2,7}}f & \dfrac{P_{2,3}P_{3,1}}{P_{2,6}P_{2,7}}c & \dfrac{P_{2,1}}{P_{2,7}}d & a & \dfrac{P_{2,3}{}^2 P_{3,1}}{P_{2,5}P_{2,6}P_{2,7}}b \\[3ex] \dfrac{P_{2,5}P_{5,1}}{P_{2,6}}f & \dfrac{P_{2,5}P_{5,1}}{P_{2,6}P_{2,7}}g & \dfrac{P_{2,5}{}^2 P_{5,1}}{P_{2,3}P_{3,1}}h & \dfrac{P_{2,5}P_{5,1}}{P_{2,1}}e & P_{2,5}b & \dfrac{P_{2,5}P_{2,6}}{P_{2,3}}c & \dfrac{P_{2,1}P_{2,5}P_{2,6}P_{2,7}}{P_{2,3}{}^2 P_{3,1}}d & a \end{bmatrix}$$

$$(18.2)$$

Counting the parameters, we see that we have seven distinct parameters giving 128 algebras; they are all commutative algebras; they have the following forms:

$$a + b\sqrt[4]{+1} + c\sqrt[2]{+1} + d\sqrt[4]{+1} + e\sqrt[2]{+1} + f\sqrt[4]{+1} + g\sqrt[2]{+1} + h\sqrt[4]{+1} \qquad 32 \;\; \textit{off} \qquad (18.3)$$

$$a + b\sqrt[4]{+1} + c\sqrt[2]{+1} + d\sqrt[4]{+1} + e\sqrt[2]{-1} + f\sqrt[4]{+1} + g\sqrt[2]{-1} + h\sqrt[4]{+1} \qquad 32 \;\; \textit{off} \qquad (18.4)$$

$$a + b\sqrt[4]{-1} + c\sqrt[2]{-1} + d\sqrt[4]{-1} + e\sqrt[2]{-1} + f\sqrt[4]{-1} + g\sqrt[2]{+1} + h\sqrt[4]{-1} \qquad 64 \;\; \textit{off} \qquad (18.5)$$

There are no geometric spaces supported by these algebras.

The computer code:

The computer code that gives these results is:

```
>  R1 := [a, b, c, d, e, f, g, h] :
>  R2 := [d, a, b, c, h, e, f, g] :
>  R3 := [c, d, a, b, g, h, e, f] :
>  R4 := [b, c, d, a, f, g, h, e] :
>  R5 := [e, f, g, h, a, b, c, d] :
>  R6 := [h, e, f, g, d, a, b, c] :
>  R7 := [g, h, e, f, c, d, a, b] :
>  R8 := [f, g, h, e, b, c, d, a] :
>
>  a := A[0] :  b := A[1] :  c := A[2] :  d := A[3] : e := A[4] :f := A[5] :
>  g := A[6] : h := A[7] :
>  AN := Matrix([R1, R2, R3, R4, R5, R6, R7, R8]) :
>  VARMAT := copy(AN) :
>  a := S[0] : b := S[1] :  c := S[2] :  d := S[3] :  e := S[4] :  f := S[5] :
>  g := S[6] : h := S[7] :
>  SY := Matrix([R1, R2, R3, R4, R5, R6, R7, R8]) :
>  a := X[0] : b := X[1] : c := X[2] : d := X[3] :  e := X[4] :  f := X[5] :
>  g := X[6] : h := X[7] :
>  VARMAT2 := Matrix([R1, R2, R3, R4, R5, R6, R7, R8]) :
>  a := B[0] : b := B[1] : c := B[2] : d := B[3] :  e := B[4] :  f := B[5] :
>  g := B[6] : h := B[7] :
>  VARMAT3 := Matrix([R1, R2, R3, R4, R5, R6, R7, R8]) :
>  a := C[0] : b := C[1] : c := C[2] : d := C[3] :  e := C[4] :  f := C[5] :
>  g := C[6] : h := C[7] :
>  VARMAT4 := Matrix([R1, R2, R3, R4, R5, R6, R7, R8]) :
>
>  for row from 1 to SIZE do
>  for col from 1 to SIZE do
>  ANDY[row, col] := P[row, col]·AN[row, col]
>  SYD[row, col] := P[row, col]·SY[row, col]
>  PARAM[row, col] := P[row, col] :
>  end do: end do:
>  for xx from 1 to SIZE do
>  P[1, xx] := 1 :
```

> $P[xx, xx] := 1$:

> **end do**:

> $VARMAT$:

>

> $P[3, 2] := \dfrac{P[2, 1]}{P[2, 3]}$: $P[4, 3] := \dfrac{P[2, 1]}{P[3, 4]}$: $P[4, 1] := P[2, 1]$: :

> $P[6, 5] := \dfrac{P[2, 1]}{P[5, 6]}$: $P[7, 6] := \dfrac{P[2, 1]}{P[6, 7]}$: $P[8, 7] := \dfrac{P[2, 1]}{P[7, 8]}$:

> $P[8, 5] := \dfrac{P[2, 1]}{P[5, 8]}$: $P[4, 2] := \dfrac{P[3, 1]}{P[2, 4]}$: $P[7, 5] := \dfrac{P[3, 1]}{P[5, 7]}$:

> $P[8, 6] := \dfrac{P[3, 1]}{P[6, 8]}$: $P[6, 2] := \dfrac{P[5, 1]}{P[2, 6]}$: $P[7, 3] := \dfrac{P[5, 1]}{P[3, 7]}$:

> $P[8, 4] := \dfrac{P[5, 1]}{P[4, 8]}$: $P[7, 2] := \dfrac{P[6, 1]}{P[2, 7]}$: $P[8, 3] := \dfrac{P[6, 1]}{P[3, 8]}$:

> $P[5, 4] := \dfrac{P[6, 1]}{P[4, 5]}$: $P[5, 2] := \dfrac{P[6, 1]}{P[2, 5]}$: $P[6, 3] := \dfrac{P[6, 1]}{P[3, 6]}$:

> $P[7, 4] := \dfrac{P[6, 1]}{P[4, 7]}$: $P[8, 1] := P[6, 1]$: $P[8, 2] := \dfrac{P[7, 1]}{P[2, 8]}$:

> $P[5, 3] := \dfrac{P[7, 1]}{P[3, 5]}$: $P[6, 4] := \dfrac{P[7, 1]}{P[4, 6]}$: # *end of diagonal*

>

> $P[3, 4] := P[2, 4]$: $P[4, 5] := \dfrac{P[2, 5] \cdot P[3, 8]}{P[2, 4]}$: $P[4, 6] := \dfrac{P[2, 6] \cdot P[3, 5]}{P[2, 4]}$:

> $P[4, 7] := \dfrac{P[2, 7] \cdot P[3, 6]}{P[2, 4]}$: $P[4, 8] := \dfrac{P[2, 8] \cdot P[3, 7]}{P[2, 4]}$:

> $P[7, 8] := \dfrac{P[2, 7] \cdot P[7, 1]}{P[6, 1]}$: $P[3, 8] := \dfrac{P[2, 7] \cdot P[3, 7] \cdot P[7, 1]}{P[2, 5] \cdot P[5, 1]}$:

> $P[6, 8] := \dfrac{P[2, 7] \cdot P[3, 1] \cdot P[6, 7] \cdot P[7, 1]}{P[2, 1] \cdot P[2, 4] \cdot P[6, 1]}$: $P[5, 8] := \dfrac{P[2, 1] \cdot P[2, 7] \cdot P[5, 7] \cdot P[7, 1]}{P[2, 3] \cdot P[3, 1] \cdot P[6, 1]}$:

> $P[2, 8] := \dfrac{P[2, 1] \cdot P[2, 4]}{P[2, 7] \cdot P[3, 6]}$: $P[2, 4] := \dfrac{P[2, 3] \cdot P[3, 1]}{P[2, 1]}$: $P[6, 7] := \dfrac{P[2, 3] \cdot P[5, 7]}{P[5, 6]}$:

> $P[5, 7] := \dfrac{P[3, 1] \cdot P[5, 6] \cdot P[6, 1]}{P[2, 7] \cdot P[3, 6] \cdot P[7, 1]}$: $P[5, 6] := \dfrac{P[2, 5] \cdot P[5, 1]}{P[6, 1]}$:

> $P[6, 1] := \dfrac{P[2, 5] \cdot P[5, 1]}{P[2, 6]}$: $P[7, 1] := \dfrac{P[2, 5] \cdot P[3, 5] \cdot P[5, 1]}{P[2, 7] \cdot P[3, 6]}$:

> $P[3, 7] := \dfrac{P[2, 3] \cdot P[3, 1] \cdot P[3, 6]}{P[2, 5] \cdot P[2, 6] \cdot P[3, 5]}$: $P[3, 6] := \dfrac{P[2, 5] \cdot P[3, 1]}{P[2, 7] \cdot P[3, 5]}$:

> $P[3, 5] := \dfrac{P[2, 3] \cdot P[3, 1]}{P[2, 6] \cdot P[2, 7]}$:

The $C_3 \times C_3$ Group

There are two order nine groups. There is the cyclic order nine group C_9 dealt with earlier, and there is the $C_3 \times C_3$ group. The $C_3 \times C_3$ group is a commutative group with four C_3 subgroups.

The Standard form Cayley table of the $C_3 \times C_3$ group is:

$$C_3 \times C_3 \sim \begin{bmatrix} a & b & c & d & e & f & g & h & i \\ c & a & b & f & d & e & i & g & h \\ b & c & a & e & f & d & h & i & g \\ g & h & i & a & b & c & d & e & f \\ i & g & h & c & a & b & f & d & e \\ h & i & g & b & c & a & e & f & d \\ d & e & f & g & h & i & a & b & c \\ f & d & e & i & g & h & c & a & b \\ e & f & d & h & i & g & b & c & a \end{bmatrix} \qquad (19.1)$$

We have nine copies of the C_3 Standard form Cayley table scattered throughout the matrix in a pattern which copies, as 3×3 blocks, copies the C_3 Standard form Cayley table.

The algebraic matrix form of the $C_3 \times C_3$ group is too large to show sensibly (see code at end of chapter).

All parameter elimination equations are linear leaving eight parameters and so the $C_3 \times C_3$ group contains 256 separate algebras. The algebras are all commutative. Those division algebras are:

$$1 \quad of \quad 1 + 8\sqrt[3]{+1} \qquad (19.2)$$

$$8 \quad off \quad 1 + 7\sqrt[3]{+1} + \sqrt[3]{-1} \qquad (19.3)$$

$$28 \quad off \quad 1 + 6\sqrt[3]{+1} + 2\sqrt[3]{-1} \qquad (19.4)$$

$$56 \quad off \quad 1 + 5\sqrt[3]{+1} + 3\sqrt[3]{-1} \qquad (19.5)$$

$$70 \quad off \quad 1 + 4\sqrt[3]{+1} + 4\sqrt[3]{-1} \qquad (19.6)$$

$$56 \quad off \quad 1 + 5\sqrt[3]{-1} + 3\sqrt[3]{+1} \qquad (19.7)$$

$$28 \quad off \quad 1 + 6\sqrt[3]{-1} + 2\sqrt[3]{+1} \qquad (19.8)$$

$$8 \quad off \quad 1 + 7\sqrt[3]{-1} + \sqrt[3]{+1} \qquad (19.9)$$

$$1 \quad of \quad 1+8\sqrt[3]{-1} \qquad\qquad (19.10)$$

There are no geometric spaces supported by the $C_3 \times C_3$ group. Since this group has C_3 subgroups, that is what we expected.

The computer code:

The computer code generating the $C_3 \times C_3$ group is:

```
>  R1 := [a, b, c, d, e, f, g, h, i] :
>  R2 := [c, a, b, f, d, e, i, g, h] :
>  R3 := [b, c, a, e, f, d, h, i, g] :
>  R4 := [g, h, i, a, b, c, d, e, f] :
>  R5 := [i, g, h, c, a, b, f, d, e] :
>  R6 := [h, i, g, b, c, a, e, f, d] :
>  R7 := [d, e, f, g, h, i, a, b, c] :
>  R8 := [f, d, e, i, g, h, c, a, b] :
>  R9 := [e, f, d, h, i, g, b, c, a] :
>
>  a := A[0] : b := A[1] : c := A[2] : d := A[3] : e := A[4] : f := A[5] :
>  g := A[6] : h := A[7] : i := A[8] :
>  AN := Matrix([R1, R2, R3, R4, R5, R6, R7, R8, R9]) :
>  VARMAT := copy(AN) :
>  a := S[0] : b := S[1] : c := S[2] : d := S[3] : e := S[4] : f := S[5] :
>  g := S[6] : h := S[7] : i := S[8] :
>  SY := Matrix([R1, R2, R3, R4, R5, R6, R7, R8, R9]) :
>  a := X[0] : b := X[1] : c := X[2] : d := X[3] : e := X[4] : f := X[5] :
>  g := X[6] : h := X[7] : i := X[8] :
>  VARMAT2 := Matrix([R1, R2, R3, R4, R5, R6, R7, R8, R9]) :
>  a := B[0] : b := B[1] : c := B[2] : d := B[3] : e := B[4] : f := B[5] :
>  g := B[6] : h := B[7] : i := B[8] :
>  VARMAT3 := Matrix([R1, R2, R3, R4, R5, R6, R7, R8, R9]) :
>  a := C[0] : b := C[1] : c := C[2] : d := C[3] : e := C[4] : f := C[5] :
>  g := C[6] : h := C[7] : i := C[8] :
>  VARMAT4 := Matrix([R1, R2, R3, R4, R5, R6, R7, R8, R9]) :
>
>  for row from 1 to SIZE do
>  for col from 1 to SIZE do
>  ANDY[row, col] := P[row, col]·AN[row, col]
>  SYD[row, col] := P[row, col]·SY[row, col]
>  PARAM[row, col] := P[row, col] :
>  end do: end do:
>  for xx from 1 to SIZE do
>  P[1, xx] := 1 :
>  P[xx, xx] := 1 :
>  end do:
>  VARMAT :
>
```

> $P[7,5] := \dfrac{P[6,1]}{P[5,7]}$: $P[6,3] := \dfrac{P[4,1]}{P[3,6]}$: $P[3,1] := P[2,1] : :$

> $P[8,4] := \dfrac{P[5,1]}{P[4,8]}$: $P[7,1] := P[4,1]$: $P[9,1] := P[5,1]$:

> $P[8,1] := P[6,1]$: $P[8,3] := \dfrac{P[5,1]}{P[3,8]}$: $P[3,2] := \dfrac{P[2,1]}{P[2,3]}$:

> $P[5,4] := \dfrac{P[2,1]}{P[4,5]}$: $P[6,5] := \dfrac{P[2,1]}{P[5,6]}$: $P[6,4] := \dfrac{P[2,1]}{P[4,6]}$:

> $P[7,4] := \dfrac{P[4,1]}{P[4,7]}$: $P[5,2] := \dfrac{P[4,1]}{P[2,5]}$: $P[8,5] := \dfrac{P[4,1]}{P[5,8]}$:

> $P[6,2] := \dfrac{P[5,1]}{P[2,6]}$: $P[4,3] := \dfrac{P[5,1]}{P[3,4]}$: $P[9,5] := \dfrac{P[5,1]}{P[5,9]}$:

> $P[4,2] := \dfrac{P[6,1]}{P[2,4]}$: $P[5,3] := \dfrac{P[6,1]}{P[3,5]}$: $P[9,4] := \dfrac{P[6,1]}{P[4,9]}$:

> $P[8,7] := \dfrac{P[2,1]}{P[7,8]}$: $P[9,8] := \dfrac{P[2,1]}{P[8,9]}$: $P[9,7] := \dfrac{P[2,1]}{P[7,9]}$:

> $P[9,6] := \dfrac{P[4,1]}{P[6,9]}$: $P[8,2] := \dfrac{P[4,1]}{P[2,8]}$: $P[9,3] := \dfrac{P[4,1]}{P[3,9]}$:

> $P[7,3] := \dfrac{P[6,1]}{P[3,7]}$: $P[9,2] := \dfrac{P[6,1]}{P[2,9]}$: $P[8,6] := \dfrac{P[6,1]}{P[6,8]}$:

> $P[7,6] := \dfrac{P[5,1]}{P[6,7]}$: $P[7,2] := \dfrac{P[5,1]}{P[2,7]}$: *# end of diagonal*

>

> $P[8,9] := \dfrac{P[2,4]\cdot P[2,9]\cdot P[4,1]}{P[2,8]\cdot P[6,1]}$: $P[7,9] := \dfrac{P[3,4]\cdot P[4,1]}{P[5,1]}$:

> $P[6,9] := \dfrac{P[4,8]\cdot P[6,1]}{P[5,1]}$: $P[5,9] := \dfrac{P[2,1]\cdot P[4,9]}{P[3,5]\cdot P[4,5]}$:

> $P[3,5] := \dfrac{P[2,1]}{P[2,6]}$: $P[6,8] := \dfrac{P[2,4]\cdot P[4,9]}{P[2,9]}$:

> $P[6,7] := \dfrac{P[4,9]\cdot P[5,1]}{P[4,1]}$: $P[4,9] := \dfrac{P[3,9]\cdot P[4,8]\cdot P[6,1]}{P[3,4]\cdot P[4,1]}$:

> $P[7,8] := \dfrac{P[2,4]\cdot P[4,1]}{P[6,1]}$: $P[5,8] := \dfrac{P[2,8]\cdot P[4,7]}{P[2,5]}$:

> $P[5,7] := \dfrac{P[4,8]\cdot P[6,1]}{P[4,1]}$: $P[3,9] := \dfrac{P[2,5]\cdot P[3,4]}{P[2,7]}$:

> $P[5,6] := \dfrac{P[2,9]\cdot P[5,1]}{P[6,1]}$: $P[2,9] := \dfrac{P[2,1]}{P[3,8]}$: $P[4,8] := \dfrac{P[2,4]\cdot P[4,1]\cdot P[4,7]}{P[4,5]\cdot P[6,1]}$:

> $P[4,5] := \dfrac{P[2,7]\cdot P[4,1]}{P[5,1]}$: $P[4,6] := \dfrac{P[3,7]\cdot P[4,1]}{P[6,1]}$: $P[3,8] := \dfrac{P[2,1]\cdot P[2,7]}{P[2,3]\cdot P[3,7]}$:

> $P[2,6] := \dfrac{P[2,1]\cdot P[2,3]}{P[2,4]\cdot P[2,5]}$: $P[3,4] := \dfrac{P[2,1]}{P[2,5]}$: $P[3,7] := \dfrac{P[2,1]}{P[2,8]}$:

> $P[3,6] := \dfrac{P[2,1]}{P[2,4]}$: $P[2,8] := \dfrac{P[2,4]\cdot P[2,7]\cdot P[4,1]^2}{P[2,5]\cdot P[5,1]\cdot P[6,1]}$: $P[2,7] := \dfrac{P[2,5]\cdot P[5,1]}{P[4,1]}$:

The $C_2 \times C_6$ Group

The $C_2 \times C_6$ group is a commutative group with three C_2 subgroups, one C_3 subgroup, one $C_2 \times C_2$ subgroup and three C_6 subgroups.

The Standard form Cayley table of the $C_2 \times C_6$ group is:

$$C_2 \times C_6 \sim \begin{bmatrix} a & b & c & d & e & f & g & h & i & j & k & l \\ f & a & b & c & d & e & l & g & h & i & j & k \\ e & f & a & b & c & d & k & l & g & h & i & j \\ d & e & f & a & b & c & j & k & l & g & h & i \\ c & d & e & f & a & b & i & j & k & l & g & h \\ b & c & d & e & f & a & h & i & j & k & l & g \\ g & h & i & j & k & l & a & b & c & d & e & f \\ l & g & h & i & j & k & f & a & b & c & d & e \\ k & l & g & h & i & j & e & f & a & b & c & d \\ j & k & l & g & h & i & d & e & f & a & b & c \\ i & j & k & l & g & h & c & d & e & f & a & b \\ h & i & j & k & l & g & b & c & d & e & f & a \end{bmatrix} \tag{19.11}$$

Since this group has subgroups which cannot hold geometric spaces, we expect no geometric spaces within this group.

The algebras within this group, with the Standard form Cayley table above, (19.11) are of the form:

$$a + b\sqrt[6]{\pm 1} + c\sqrt[3]{\pm 1} + d\sqrt[2]{\pm 1} + e\sqrt[3]{\pm 1} + f\sqrt[6]{\pm 1} + g\sqrt[2]{\pm 1} + h\sqrt[6]{\pm 1} + i\sqrt[6]{\pm 1} + j\sqrt[2]{\pm 1} + k\sqrt[6]{\pm 1} + l\sqrt[6]{\pm 1}$$
$$(19.12)$$

The parameter elimination code of this order twelve group leaves eleven free parameters. The code ends with a single quadratic elimination equation.

When we take the positive root of the quadratic parameter elimination equation, we get six non-isomorphic commutative algebras:

$$128 \quad off \quad 1 + 6\sqrt[6]{+1} + 2\sqrt[3]{-1} + 3\sqrt[2]{+1} \quad Comm \tag{19.13}$$

$$128 \quad off \quad 1 + 6\sqrt[6]{+1} + 2\sqrt[3]{+1} + 3\sqrt[2]{+1} \quad Comm \tag{19.14}$$

$$256 \quad off \quad 1 + 6\sqrt[6]{+1} + \sqrt[3]{+1} + 3\sqrt[2]{+1} + \sqrt[3]{-1} \quad Comm \tag{19.15}$$

$$384 \quad off \quad 1 + 2\sqrt[6]{+1} + 2\sqrt[3]{+1} + \sqrt[2]{+1} + 2\sqrt[3]{-1} + 4\sqrt[6]{-1} \quad Comm \tag{19.16}$$

$$384 \quad off \quad 1 + 4\sqrt[6]{-1} + 2\sqrt[3]{-1} + 2\sqrt[2]{-1} + \sqrt[3]{+1} + 2\sqrt[6]{+1} \quad Comm \tag{19.17}$$

$$768 \quad off \quad 1 + 2\sqrt[6]{+1} + \sqrt[3]{+1} + \sqrt[2]{+1} + \sqrt[3]{-1} + 2\sqrt[2]{-1} + 4\sqrt[6]{-1} \quad Comm \tag{19.18}$$

100

When we take the negative root of the quadratic parameter elimination equation, we get six non-isomorphic non-commutative algebras:

$$128 \quad off \quad 1+6\sqrt[6]{-1}+2\sqrt[3]{-1}+3\sqrt[2]{-1} \qquad Non-Comm \tag{19.19}$$

$$128 \quad off \quad 1+6\sqrt[6]{-1}+2\sqrt[3]{+1}+3\sqrt[2]{-1} \qquad Non-Comm \tag{19.20}$$

$$256 \quad off \quad 1+6\sqrt[6]{-1}+\sqrt[3]{-1}+3\sqrt[2]{-1}+\sqrt[3]{+1} \qquad Non-Comm \tag{19.21}$$

$$384 \quad off \quad 1+4\sqrt[6]{+1}+2\sqrt[3]{+1}+2\sqrt[2]{+1}+2\sqrt[6]{-1}+\sqrt[2]{-1} \qquad Non-Comm \tag{19.22}$$

$$384 \quad off \quad 1+2\sqrt[6]{-1}+2\sqrt[3]{-1}+\sqrt[3]{-1}+2\sqrt[2]{+1}+4\sqrt[6]{+1} \qquad Non-Comm \tag{19.23}$$

$$768 \quad off \quad 1+4\sqrt[6]{+1}+\sqrt[3]{+1}+2\sqrt[2]{+1}+\sqrt[3]{-1}+2\sqrt[6]{-1}+\sqrt[2]{-1} \qquad Non-Comm \tag{19.24}$$

We see that, like the $C_2 \times C_2$ group, the quadratic parameter elimination equation has led to a set of non-commutative algebras.

These non-commutative algebras are chiral algebra; by this we mean that the non-commutative pairs of variables have a zero anti-commutator.

The computer code:

The computer code of the order twelve $C_2 \times C_6$ group is:

```
> R1 := [a, b, c, d, e, f, g, h, i, j, k, l] :
> R2 := [f, a, b, c, d, e, l, g, h, i, j, k] :
> R3 := [e, f, a, b, c, d, k, l, g, h, i, j] :
> R4 := [d, e, f, a, b, c, j, k, l, g, h, i] :
> R5 := [c, d, e, f, a, b, i, j, k, l, g, h] :
> R6 := [b, c, d, e, f, a, h, i, j, k, l, g] :
> R7 := [g, h, i, j, k, l, a, b, c, d, e, f] :
> R8 := [l, g, h, i, j, k, f, a, b, c, d, e] :
> R9 := [k, l, g, h, i, j, e, f, a, b, c, d] :
> R10 := [j, k, l, g, h, i, d, e, f, a, b, c] :
> R11 := [i, j, k, l, g, h, c, d, e, f, a, b] :
> R12 := [h, i, j, k, l, g, b, c, d, e, f, a] :
>
> a := A[0] : b := A[1] : c := A[2] : d := A[3] : e := A[4] : f := A[5] :
> g := A[6] : h := A[7] : i := A[8] : j := A[9] : k := A[10] : l := A[11] :
> AN := Matrix([R1, R2, R3, R4, R5, R6, R7, R8, R9, R10, R11, R12]) :
> VARMAT := copy(AN) :
> a := S[0] : b := S[1] : c := S[2] : d := S[3] : e := S[4] : f := S[5] :
> g := S[6] : h := S[7] : i := S[8] : j := S[9] : k := S[10] : l := S[11] :
> SY := Matrix([R1, R2, R3, R4, R5, R6, R7, R8, R9, R10, R11, R12]) :
> a := X[0] : b := X[1] : c := X[2] : d := X[3] : e := X[4] : f := X[5] :
> g := X[6] : h := X[7] : i := X[8] : j := X[9] : k := X[10] : l := X[11] :
> VARMAT2 := Matrix([R1, R2, R3, R4, R5, R6, R7, R8, R9, R10, R11, R12]) :
```

> $a := B[0] : b := B[1] : c := B[2] : d := B[3] : e := B[4] : f := B[5] :$
> $g := B[6] : h := B[7] : i := B[8] : j := B[9] : k := B[10] : l := B[11] :$
> $VARMAT3 := Matrix([R1, R2, R3, R4, R5, R6, R7, R8, R9, R10, R11, R12]) :$
> $a := C[0] : b := C[1] : c := C[2] : d := C[3] : e := C[4] : f := C[5] :$
> $g := C[6] : h := C[7] : i := C[8] : j := C[9] : k := C[10] : l := C[11] :$
> $VARMAT4 := Matrix([R1, R2, R3, R4, R5, R6, R7, R8, R9, R10, R11, R12]) :$
>
> **for** *row* **from** 1 **to** *SIZE* **do**
> **for** *col* **from** 1 **to** *SIZE* **do**
> $ANDY[row, col] := P[row, col] \cdot AN[row, col]$
> $SYD[row, col] := P[row, col] \cdot SY[row, col]$
> $PARAM[row, col] := P[row, col] :$
> **end do: end do:**
> **for** *xx* **from** 1 **to** *SIZE* **do**
> $P[1, xx] := 1 :$
> $P[xx, xx] := 1 :$
> **end do:**
>
> $P[3, 2] := \dfrac{P[2, 1]}{P[2, 3]} : P[4, 3] := \dfrac{P[2, 1]}{P[3, 4]} : P[6, 1] := P[2, 1] :$
>
> $P[5, 4] := \dfrac{P[2, 1]}{P[4, 5]} : P[6, 5] := \dfrac{P[2, 1]}{P[5, 6]} : P[8, 7] := \dfrac{P[2, 1]}{P[7, 8]} :$
>
> $P[11, 10] := \dfrac{P[2, 1]}{P[10, 11]} : P[9, 8] := \dfrac{P[2, 1]}{P[8, 9]} : P[12, 11] := \dfrac{P[2, 1]}{P[11, 12]} :$
>
> $P[10, 9] := \dfrac{P[2, 1]}{P[9, 10]} : P[12, 7] := \dfrac{P[2, 1]}{P[7, 12]} : P[4, 2] := \dfrac{P[3, 1]}{P[2, 4]} :$
>
> $P[5, 3] := \dfrac{P[3, 1]}{P[3, 5]} : P[6, 4] := \dfrac{P[3, 1]}{P[4, 6]} : P[5, 1] := P[3, 1] :$
>
> $P[6, 2] := \dfrac{P[3, 1]}{P[2, 6]} : P[9, 7] := \dfrac{P[3, 1]}{P[7, 9]} : P[10, 8] := \dfrac{P[3, 1]}{P[8, 10]} :$
>
> $P[11, 9] := \dfrac{P[3, 1]}{P[9, 11]} : P[12, 10] := \dfrac{P[3, 1]}{P[10, 12]} : P[11, 7] := \dfrac{P[3, 1]}{P[7, 11]} :$
>
> $P[12, 8] := \dfrac{P[3, 1]}{P[8, 12]} : P[9, 7] := \dfrac{P[3, 1]}{P[7, 9]} : P[10, 8] := \dfrac{P[3, 1]}{P[8, 10]} :$
>
> $P[5, 2] := \dfrac{P[4, 1]}{P[2, 5]} : P[6, 3] := \dfrac{P[4, 1]}{P[3, 6]} : P[11, 8] := \dfrac{P[4, 1]}{P[8, 11]} :$
>
> $P[12, 9] := \dfrac{P[4, 1]}{P[9, 12]} : P[10, 7] := \dfrac{P[4, 1]}{P[7, 10]} : P[10, 4] := \dfrac{P[7, 1]}{P[4, 10]} :$
>
> $P[8, 2] := \dfrac{P[7, 1]}{P[2, 8]} : P[9, 3] := \dfrac{P[7, 1]}{P[3, 9]} : P[11, 5] := \dfrac{P[7, 1]}{P[5, 11]} :$
>
> $P[12, 6] := \dfrac{P[7, 1]}{P[6, 12]} : P[12, 1] := P[8, 1] :$
>
> $P[9, 2] := \dfrac{P[8, 1]}{P[2, 9]} : P[7, 6] := \dfrac{P[8, 1]}{P[6, 7]} : P[11, 4] := \dfrac{P[8, 1]}{P[4, 11]} :$
>
> $P[12, 5] := \dfrac{P[8, 1]}{P[5, 12]} : P[10, 5] := \dfrac{P[8, 1]}{P[5, 10]} : P[10, 3] := \dfrac{P[8, 1]}{P[3, 10]} :$
>
> $P[8, 3] := \dfrac{P[8, 1]}{P[3, 8]} : P[9, 4] := \dfrac{P[8, 1]}{P[4, 9]} : P[11, 6] := \dfrac{P[8, 1]}{P[6, 11]} :$
>
> $P[7, 2] := \dfrac{P[8, 1]}{P[2, 7]} : P[10, 2] := \dfrac{P[9, 1]}{P[2, 10]} : P[11, 3] := \dfrac{P[9, 1]}{P[3, 11]} :$
>
> $P[12, 4] := \dfrac{P[9, 1]}{P[4, 12]} : P[7, 5] := \dfrac{P[9, 1]}{P[5, 7]} : P[8, 6] := \dfrac{P[9, 1]}{P[6, 8]} :$

> $P[7,3] := \dfrac{P[9,1]}{P[3,7]}$: $P[8,4] := \dfrac{P[9,1]}{P[4,8]}$: $P[9,5] := \dfrac{P[9,1]}{P[5,9]}$:

> $P[10,6] := \dfrac{P[9,1]}{P[6,10]}$: $P[12,2] := \dfrac{P[9,1]}{P[2,12]}$: $P[11,1] := P[9,1]$:

> $P[11,2] := \dfrac{P[10,1]}{P[2,11]}$: $P[12,3] := \dfrac{P[10,1]}{P[3,12]}$: $P[7,4] := \dfrac{P[10,1]}{P[4,7]}$:

> $P[8,5] := \dfrac{P[10,1]}{P[5,8]}$: $P[9,6] := \dfrac{P[10,1]}{P[6,9]}$: # *Leading diagonal done*

>

> $P[11,12] := \dfrac{P[2,9] \cdot P[9,1]}{P[8,1]}$: $P[10,12] := \dfrac{P[3,10] \cdot P[10,1]}{P[8,1]}$:

> $P[9,12] := \dfrac{P[4,11] \cdot P[9,1]}{P[8,1]}$: $P[8,12] := P[5,12]$:

> $P[7,12] := \dfrac{P[5,12] \cdot P[7,8]}{P[5,6]}$: $P[5,12] := \dfrac{P[4,12] \cdot P[5,9]}{P[4,8]}$:

> $P[5,10] := \dfrac{P[5,8] \cdot P[8,1]}{P[10,1]}$: $P[6,12] := \dfrac{P[2,1]}{P[2,7]}$: $P[4,12] := \dfrac{P[4,1]}{P[4,9]}$:

> $P[9,11] := \dfrac{P[3,1]}{P[5,9]}$: $P[5,9] := \dfrac{P[2,9] \cdot P[4,8]}{P[4,5]}$:

> $P[10,11] := \dfrac{P[2,10] \cdot P[10,1]}{P[9,1]}$: $P[8,11] := \dfrac{P[2,5] \cdot P[7,11]}{P[7,8]}$:

> $P[7,11] := \dfrac{P[5,7] \cdot P[7,1]}{P[9,1]}$: $P[6,11] := \dfrac{P[2,1]}{P[2,12]}$:

> $P[5,11] := \dfrac{P[5,7] \cdot P[7,1]}{P[9,1]}$: $P[7,9] := \dfrac{P[3,1]}{P[5,7]}$: $P[4,11] := \dfrac{P[4,1]}{P[4,8]}$:

> $P[9,10] := \dfrac{P[2,11] \cdot P[9,1]}{P[10,1]}$: $P[5,8] := \dfrac{P[2,11] \cdot P[4,8]}{P[4,5]}$:

> $P[8,10] := P[3,10]$: $P[3,11] := \dfrac{P[3,10] \cdot P[6,8]}{P[6,10]}$:

> $P[3,12] := \dfrac{P[3,10] \cdot P[10,1]}{P[8,1]}$: $P[7,10] := \dfrac{P[3,10] \cdot P[7,8]}{P[2,4]}$:

> $P[6,10] := \dfrac{P[3,7] \cdot P[4,10]}{P[4,6]}$: $P[4,10] := \dfrac{P[4,1]}{P[4,7]}$:

> $P[8,9] := \dfrac{P[2,12] \cdot P[8,1]}{P[9,1]}$:

> $P[6,9] := \dfrac{P[2,1]}{P[2,10]}$: $P[4,9] := \dfrac{P[2,9] \cdot P[4,8]}{P[2,12]}$:

> $P[4,5] := \dfrac{P[2,4] \cdot P[4,1]}{P[3,1]}$:

> $P[3,9] := \dfrac{P[3,1]}{P[5,7]}$: $P[5,6] := P[2,6]$: $P[7,8] := \dfrac{P[2,7] \cdot P[7,1]}{P[8,1]}$:

> $P[6,8] := \dfrac{P[2,1]}{P[2,9]}$: $P[4,6] := \dfrac{P[2,4] \cdot P[4,1]}{P[2,1]}$:

> $P[4,8] := \dfrac{P[2,12] \cdot P[3,8]}{P[3,4]}$: $P[3,4] := P[2,4]$:

> $P[3,6] := \dfrac{P[2,4] \cdot P[4,1]}{P[2,1]}$: $P[3,8] := \dfrac{P[2,7] \cdot P[2,8]}{P[2,3]}$:

> $P[6,7] := \dfrac{P[2,1]}{P[2,8]}$: $P[2,12] := \dfrac{P[2,3] \cdot P[3,7]}{P[2,7]}$:

> $P[4,7] := \dfrac{P[2,11] \cdot P[3,7]}{P[2,4]}$: $P[2,6] := \dfrac{P[2,3] \cdot P[3,1]}{P[2,1]}$:

> $P[5,7] := \dfrac{P[3,1] \cdot P[9,1]}{P[3,7] \cdot P[7,1]}$: $P[2,5] := \dfrac{P[2,4] \cdot P[4,1]}{P[3,1]}$:

> $P[3,10] := \dfrac{P[2,9] \cdot P[2,10]}{P[2,3]}$: $P[2,9] := \dfrac{P[2,3] \cdot P[3,7] \cdot P[7,1]}{P[2,8] \cdot P[9,1]}$:

> $P[2,11] := \dfrac{P[2,4]^2 \cdot P[4,1] \cdot P[9,1]}{P[2,10] \cdot P[3,7]^2 \cdot P[7,1]}$:

> $P[10,1] := \dfrac{P[2,4]^2 \cdot P[2,8] \cdot P[4,1] \cdot P[8,1] \cdot P[9,1]^2}{P[2,7] \cdot P[2,10]^2 \cdot P[3,7]^2 \cdot P[7,1]^2}$:

> $P[3,5] := \dfrac{P[2,4]^2 \cdot P[4,1]}{P[2,3] \cdot P[3,1]}$:

> $PM8 := 0 :$ *# We set PM8 to either 0 or 1 to get the two different sets of algebras*

> $P[8,1] := \dfrac{(-1)^{PM8} \cdot P[2,7] \cdot P[7,1]}{P[2,8]}$:

SECTION IV – The Non-Abelian Groups

Chapter 20

The Order Six Symmetric Group

The order six symmetric group S_3 is isomorphic to and also known as the dihedral group D_3. S_3 is a non-commutative group; it is the lowest order non-commutative group. S_3 has three C_2 subgroups and one C_3 subgroup. The Standard form Cayley table of S_3 is:

$$S_3 = \begin{bmatrix} a & b & c & d & e & f \\ c & a & b & e & f & d \\ b & c & a & f & d & e \\ d & e & f & a & b & c \\ e & f & d & c & a & b \\ f & d & e & b & c & a \end{bmatrix} \tag{20.1}$$

We can see the C_3 Standard Cayley table appearing twice on the leading diagonal and twice as a reflection on the off diagonal. When we calculate the parameters to form the algebraic matrix form, the non-commutativity is of no concern because we seek only multiplicative closure. The algebraic matrix form is:

$$S_3 = \begin{bmatrix}
a & b & c & d & e & f \\[6pt]
P_{2,1}c & a & \dfrac{(P_{2,1})^2 (P_{2,4})^3}{(P_{3,4})^3 (P_{5,6})^3}b & P_{2,4}e & \dfrac{(P_{2,1})^2 (P_{2,4})^2}{(P_{3,4})^2 (P_{5,6})^3}f & \dfrac{P_{2,1}}{P_{3,4}}d \\[12pt]
P_{2,1}b & \dfrac{(P_{3,4})^3 (P_{5,6})^3}{P_{2,1}(P_{2,4})^3}c & a & P_{3,4}f & \dfrac{P_{2,1}}{P_{2,4}}d & \dfrac{(P_{3,4})^3 (P_{5,6})^3}{P_{2,1}(P_{2,4})^2}e \\[12pt]
P_{4,1}d & \dfrac{(P_{3,4})^2 P_{4,1}(P_{5,6})^2}{(P_{2,1})^2 P_{2,4}}e & \dfrac{(P_{2,4})^2 P_{4,1}}{P_{3,4}(P_{5,6})^2}f & a & \dfrac{(P_{2,1})^2 P_{2,4}}{(P_{3,4})^2 (P_{5,6})^2}b & \dfrac{P_{3,4}(P_{5,6})^2}{(P_{2,4})^2}c \\[12pt]
\dfrac{(P_{3,4})^2 P_{4,1}(P_{5,6})^2}{(P_{2,1})^2}e & \dfrac{(P_{3,4})^2 P_{4,1}P_{5,6}}{(P_{2,1})^2}f & \dfrac{P_{2,4}P_{4,1}}{P_{2,1}}d & \dfrac{(P_{3,4})^2 (P_{5,6})^2}{P_{2,1}P_{2,4}}c & a & P_{5,6}b \\[12pt]
\dfrac{(P_{2,4})^2 P_{4,1}}{(P_{5,6})^2}f & \dfrac{P_{4,1}P_{3,4}}{P_{2,1}}d & \dfrac{(P_{2,4})^2 P_{4,1}}{P_{2,1}P_{5,6}}e & \dfrac{P_{2,1}(P_{2,4})^2}{P_{3,4}(P_{5,6})^2}b & \dfrac{P_{2,1}}{P_{5,6}}c & a
\end{bmatrix}$$

$$(20.2)$$

The parameter elimination equations are all linear (no quadratic) and so we have a unique algebraic matrix form. This matrix, (20.2), is non-commutative for all sets of parameters, and so all the algebras which it represents are non-commutative. A typical commutator is of the form:

$$\begin{bmatrix} 0 & e-d & e-d & b+c & -b-c & 0 \\ e-d & 0 & e-d & b+c & 0 & b+c \\ d-e & d-e & 0 & 0 & b+c & b+c \\ -b-c & -b-c & 0 & 0 & d-e & d-e \\ b+c & 0 & -b-c & e-d & 0 & e-d \\ 0 & -b-c & -b-c & e-d & d-c & 0 \end{bmatrix} \tag{20.3}$$

There are 32 division algebras of the S_3 group (each formed by taking the exponential of (20.2) for a given set of parameters set to ± 1); of these, there are six algebraically distinct types of division algebra. We have:

$$a + b\sqrt[3]{-1} + c\sqrt[3]{-1} + d\sqrt[2]{-1} + e\sqrt[2]{-1} + f\sqrt[2]{-1} \qquad 4 \; off \tag{20.4}$$

$$a + b\sqrt[3]{-1} + c\sqrt[3]{-1} + d\sqrt[2]{+1} + e\sqrt[2]{+1} + f\sqrt[2]{+1} \qquad 4 \; off \tag{20.5}$$

$$a + b\sqrt[3]{-1} + c\sqrt[3]{+1} + d\sqrt[2]{-1} + e\sqrt[2]{-1} + f\sqrt[2]{-1} \qquad 8 \; off \tag{20.6}$$

$$a + b\sqrt[3]{-1} + c\sqrt[3]{+1} + d\sqrt[2]{+1} + e\sqrt[2]{+1} + f\sqrt[2]{+1} \qquad 8 \; off \tag{20.7}$$

$$a + b\sqrt[3]{+1} + c\sqrt[3]{+1} + d\sqrt[2]{-1} + e\sqrt[2]{-1} + f\sqrt[2]{-1} \qquad 4 \; off \tag{20.8}$$

$$a + b\sqrt[3]{+1} + c\sqrt[3]{+1} + d\sqrt[2]{+1} + e\sqrt[2]{+1} + f\sqrt[2]{+1} \qquad 4 \; off \tag{20.9}$$

The S_3 group has a subgroup of odd prime order, C_3; therefore, we have no geometric spaces emerging from the S_3 group. Computer calculations verify this.

Note:

It took your author three attempts to calculate the above algebraic matrix form of this group. The failure of the first two attempts might be down to human error, but your author could not trace this error. It seems that choosing to eliminate parameters in a particular order or in a particular way effects the outcome of the calculation. This is not understood. Once we have a multiplicatively closed algebraic matrix form, we definitely have a set of algebras, but we wonder why some calculation routes seem not to produce the requisite algebraic matrix form whereas other calculation routes do produce the algebraic matrix form.

The difficulties are common to most larger groups. Perhaps it is all human error.

The difficulty in calculating the algebraic matrix form of larger groups would probably be eased if we had a deeper group theoretic understanding of how the parameters of the algebraic matrix form are distributed throughout the matrix.

The code for the order six symmetric algebraic matrix form:

The Maple 17 code used to form the algebraic S_3 matrix form is:

```
>  R1 := [a, b, c, d, e, f] :
>  R2 := [c, a, b, e, f, d] :
>  R3 := [b, c, a, f, d, e] :
>  R4 := [d, e, f, a, b, c] :
>  R5 := [e, f, d, c, a, b] :
>  R6 := [f, d, e, b, c, a] :
>
>  a := A[0] : b := A[1] : c := A[2] : d := A[3] : e := A[4] : f := A[5] :
>  AN := Matrix([R1, R2, R3, R4, R5, R6]) :
>  VARMAT := copy(AN) :
>  a := S[0] : b := S[1] : c := S[2] : d := S[3] : e := S[4] : f := S[5] :
>  SY := Matrix([R1, R2, R3, R4, R5, R6]) :
>
>  a := X[0] : b := X[1] : c := X[2] : d := X[3] : e := X[4] : f := X[5] :
>  VARMAT2 := Matrix([R1, R2, R3, R4, R5, R6]) :
>  a := B[0] : b := B[1] : c := B[2] : d := B[3] : e := B[4] : f := B[5] :
>  VARMAT3 := Matrix([R1, R2, R3, R4, R5, R6]) :
>  a := C[0] : b := C[1] : c := C[2] : d := C[3] : e := C[4] : f := C[5] :
>  VARMAT4 := Matrix([R1, R2, R3, R4, R5, R6]) :
>
>  for row from 1 to SIZE do
>  for col from 1 to SIZE do
>  ANDY[row, col] := P[row, col]·AN[row, col]
>  SYD[row, col] := P[row, col]·SY[row, col]
>  PARAM[row, col] := P[row, col] :
>  end do: end do:
>  for xx from 1 to SIZE do
>  P[1, xx] := 1 :
>  P[xx, xx] := 1 :
>  end do:
>  #VARMAT;
>
```

$$> \quad P[3,2] := \frac{P[2,1]}{P[2,3]} : P[3,1] := P[2,1] : P[6,2] := \frac{P[4,1]}{P[2,6]} : P[4,2] := \frac{P[5,1]}{P[2,4]} :$$

$$> \quad P[5,2] := \frac{P[6,1]}{P[2,5]} : P[5,3] := \frac{P[4,1]}{P[3,5]} : P[6,3] := \frac{P[5,1]}{P[3,6]} : P[4,3] := \frac{P[6,1]}{P[3,4]} :$$

$$> \quad P[5,4] := \frac{P[2,1]}{P[4,5]} : P[6,4] := \frac{P[2,1]}{P[4,6]} : P[6,5] := \frac{P[2,1]}{P[5,6]} :$$

$$> \quad P[4,6] := \frac{P[2,1]·P[2,6]}{P[2,5]·P[5,6]} : P[4,5] := \frac{P[2,5]·P[5,6]}{P[2,4]} : P[3,6] := \frac{P[2,1]}{P[2,5]} :$$

$$> \quad P[3,5] := \frac{P[2,6]·P[2,5]}{P[2,3]} : P[2,6] := \frac{P[2,1]}{P[3,4]} : P[2,5] := \frac{P[2,3]·P[3,4]}{P[2,4]} :$$

$$> \quad P[5,1] := \frac{P[6,1]·P[2,4]·P[5,6]}{P[2,3]·P[3,4]} : P[6,1] := \frac{(P[2,4])^2·P[4,1]}{(P[5,6])^2} :$$

> $P[2,3] := \dfrac{(P[2,1])^2 \cdot (P[2,4])^3}{(P[3,4])^3 \cdot (P[5,6])^3} \ :$

Chapter 21

The Dicyclic Groups

The order eight dicyclic group is also called the order eight quaternion group. The order eight quaternion group is so named because each of its elements corresponds to an element of the quaternions. Within the Standard form Cayley table below, (21.2), the correspondence is:

$$a = 1, \quad b = i, \quad c = -1, \quad d = -i, \quad e = j, \quad f = k, \quad g = -j, \quad h = -k \tag{21.1}$$

The quaternion group is a non-commutative group. The subgroups are one C_2 and three C_4. Every subgroup of the quaternion group is normal; a group with this property is called a Hamiltonian group. The quaternion group is the lowest order dicyclic group. The dicyclic groups are of order $4n : n > 1$. The factor group of the quaternion group, $\dfrac{Q}{\{\pm 1\}}$ is isomorphic to the group $C_2 \times C_2$. The inner automorphism group of the quaternion group is isomorphic to the group $C_2 \times C_2$. The reader will recall that the group $C_2 \times C_2$ is the only group so far that supports geometric spaces.

The Standard form Cayley table of the quaternion group is:

$$Q \sim \begin{bmatrix} a & b & c & d & e & f & g & h \\ d & a & b & c & h & e & f & g \\ c & d & a & b & g & h & e & f \\ b & c & d & a & f & g & h & e \\ g & f & e & h & a & d & c & b \\ h & g & f & e & b & a & d & c \\ e & h & g & f & c & b & a & d \\ f & e & h & g & d & c & b & a \end{bmatrix} \tag{21.2}$$

The subgroups are: $\{a,c\}$, $\{a,b,c,d\}$, $\{a,c,e,g\}$, $\{a,c,f,h\}$.

The algebraic matrix form is too large to sensibly display in this book in one piece. However, the quaternion group has a very interesting property which can be seen only by knowing the algebraic matrix form. We therefore present the quaternion group's algebraic matrix form in two separate bits. Those bits are:

$$Q_{Left} \sim \begin{bmatrix} a & b & c & d \\[2mm] P_{2,1}d & a & P_{2,3}b & \dfrac{P_{2,3}\left(P_{2,6}\right)^2\left(P_{3,5}\right)^2\left(P_{6,1}\right)^2}{P_{2,1}\left(P_{2,5}\right)^2\left(P_{5,1}\right)^2}c \\[4mm] \dfrac{\left(P_{2,6}\right)^2\left(P_{3,5}\right)^2\left(P_{6,1}\right)^2}{\left(P_{2,5}\right)^2\left(P_{5,1}\right)^2}c & \dfrac{P_{2,1}}{P_{2,3}}d & a & \dfrac{P_{2,3}\left(P_{2,6}\right)^2\left(P_{3,5}\right)^2\left(P_{6,1}\right)^2}{P_{2,1}\left(P_{2,5}\right)^2\left(P_{5,1}\right)^2}b \\[4mm] P_{2,1}b & \dfrac{P_{2,1}}{P_{2,3}}c & \dfrac{\left(P_{2,1}\right)^2\left(P_{2,5}\right)^2\left(P_{5,1}\right)^2}{P_{2,3}\left(P_{2,6}\right)^2\left(P_{3,5}\right)^2\left(P_{6,1}\right)^2}d & a \\[4mm] P_{5,1}g & \dfrac{P_{6,1}}{P_{2,5}}f & \dfrac{P_{5,1}}{P_{3,5}}e & \dfrac{P_{2,6}P_{6,1}}{P_{2,1}}h \\[4mm] P_{6,1}h & \dfrac{P_{5,1}}{P_{2,6}}g & \dfrac{P_{2,3}P_{6,1}}{P_{2,5}P_{2,6}}f & \dfrac{P_{2,3}P_{2,6}P_{3,5}\left(P_{6,1}\right)^2}{P_{2,1}\left(P_{2,5}\right)^2 P_{5,1}}e \\[4mm] P_{5,1}e & \dfrac{\left(P_{2,5}\right)^2\left(P_{5,1}\right)^2}{P_{2,3}P_{2,6}P_{3,5}P_{6,1}}h & \dfrac{\left(P_{2,5}\right)^2\left(P_{5,1}\right)^2}{\left(P_{2,6}\right)^2 P_{3,5}\left(P_{6,1}\right)^2}g & \dfrac{P_{2,3}P_{3,5}P_{6,1}}{P_{2,1}P_{2,5}}f \\[4mm] P_{6,1}f & \dfrac{P_{2,5}P_{5,1}}{P_{2,3}P_{3,5}}e & \dfrac{\left(P_{2,5}\right)^3\left(P_{5,1}\right)^2}{P_{2,3}P_{2,6}\left(P_{3,5}\right)^2 P_{6,1}}h & \dfrac{P_{2,5}P_{5,1}}{P_{2,1}}g \end{bmatrix}$$

(21.3)

$$Q_{Right} \sim \begin{bmatrix} e & f & g & h \\[2mm] P_{2,5}h & P_{2,6}e & \dfrac{P_{2,3}P_{2,6}P_{3,5}\left(P_{6,1}\right)^2}{\left(P_{2,5}\right)^2\left(P_{5,1}\right)^2}f & \dfrac{P_{2,3}P_{3,5}}{P_{2,5}}g \\[4mm] P_{3,5}g & \dfrac{P_{2,5}P_{2,6}}{P_{2,3}}h & \dfrac{\left(P_{2,6}\right)^2 P_{3,5}\left(P_{6,1}\right)^2}{\left(P_{2,5}\right)^2\left(P_{5,1}\right)^2}e & \dfrac{P_{2,3}P_{2,6}\left(P_{3,5}\right)^2\left(P_{6,1}\right)^2}{\left(P_{2,5}\right)^3\left(P_{5,1}\right)^2}f \\[4mm] \dfrac{P_{2,1}}{P_{2,6}}f & \dfrac{P_{2,1}\left(P_{2,5}\right)^2\left(P_{5,1}\right)^2}{P_{2,3}P_{2,6}P_{3,5}\left(P_{6,1}\right)^2}g & \dfrac{P_{2,1}P_{2,5}}{P_{2,3}P_{3,5}}h & \dfrac{P_{2,1}}{P_{2,5}}e \\[4mm] a & \dfrac{P_{2,1}P_{2,5}P_{5,1}}{P_{2,3}P_{3,5}P_{6,1}}d & \dfrac{\left(P_{2,6}\right)^2 P_{3,5}\left(P_{6,1}\right)^2}{\left(P_{2,5}\right)^2\left(P_{5,1}\right)^2}c & \dfrac{P_{2,3}P_{2,6}P_{3,5}P_{6,1}}{\left(P_{2,5}\right)^2 P_{5,1}}b \\[4mm] \dfrac{P_{2,3}P_{3,5}P_{6,1}}{P_{2,5}P_{5,1}}b & a & \dfrac{P_{2,1}P_{6,1}}{P_{2,5}P_{5,1}}d & \dfrac{P_{2,3}P_{2,6}\left(P_{3,5}\right)^2\left(P_{6,1}\right)^2}{\left(P_{2,5}\right)^3\left(P_{5,1}\right)^2}c \\[4mm] P_{3,5}c & \dfrac{P_{2,5}P_{5,1}}{P_{6,1}}b & a & \dfrac{P_{2,1}P_{5,1}}{P_{2,6}P_{6,1}}d \\[4mm] \dfrac{P_{2,1}\left(P_{2,5}\right)^2 P_{5,1}}{P_{2,3}P_{2,6}P_{3,5}P_{6,1}}d & \dfrac{P_{2,5}P_{2,6}}{P_{2,3}}c & \dfrac{P_{2,6}P_{6,1}}{P_{5,1}}b & a \end{bmatrix}$$

(21.4)

110

All the parameter elimination equations are linear (only one solution). There are 128 separate algebras, but, remarkably, they are all isomorphic copies of the same algebra. That algebra is of the form:

$$a + b\sqrt[4]{+1} + c\sqrt[2]{+1} + d\sqrt[4]{+1} + e\sqrt[4]{+1} + f\sqrt[4]{+1} + g\sqrt[4]{+1} + h\sqrt[4]{+1} \tag{21.5}$$

Which, of course, reflects the subgroups of the quaternion group. None-the-less it is remarkable that we have only one algebra from this group. The explanation for this is the very special nature of the algebraic matrix form given in two pieces above, (21.3) & (21.4).

The algebra is non-commutative.

Let us set all variables to zero except, say, the c variable and square the algebraic matrix form. The c variable is a square root, and so the squared algebraic matrix form will be zero everywhere except on the leading diagonal. Those leading diagonal elements are all equal and are:

$$\frac{\left(P_{2,6}\right)^2 \left(P_{3,5}\right)^2 \left(P_{6,1}\right)^2}{\left(P_{2,5}\right)^2 \left(P_{5,1}\right)^2} c^2 \tag{21.6}$$

Similarly, if we do the same with another variable, say g, and take the fourth power of the algebraic matrix form, we get:

$$\frac{\left(P_{2,5}\right)^2 \left(P_{5,1}\right)^2}{\left(P_{2,6}\right)^2 \left(P_{6,1}\right)^2} g^4 \tag{21.7}$$

We get similar results for every variable. The point is that, no matter whether the parameters are plus unity or minus unity, we always get plus unity for the root because every parameter is squared. Thus, of the 128 different permutations of the parameters as ± 1 every permutation gives seven roots of plus unity, and so all 128 algebras are the same.

By computer calculation, we verify that there are no emergent distance functions which support a geometric space. Since there is a C_4 subgroup in this group, that does not surprise us.

The code of the quaternion group:
The Maple 17 code used to produce the quaternion group algebraic matrix form is:

```
> R1 := [a, b, c, d, e, f, g, h] :
> R2 := [d, a, b, c, h, e, f, g] :
> R3 := [c, d, a, b, g, h, e, f] :
> R4 := [b, c, d, a, f, g, h, e] :
> R5 := [g, f, e, h, a, d, c, b] :
> R6 := [h, g, f, e, b, a, d, c] :
> R7 := [e, h, g, f, c, b, a, d] :
> R8 := [f, e, h, g, d, c, b, a] :
>
> a := A[0] : b := A[1] : c := A[2] : d := A[3] : e := A[4] : f := A[5] :
> g := A[6] : h := A[7] :
```

> $AN := Matrix([R1, R2, R3, R4, R5, R6, R7, R8])$:
> $VARMAT := copy(AN)$:
> $a := S[0] : b := S[1] : c := S[2] : d := S[3] : e := S[4] : f := S[5]$:
> $g := S[6] : h := S[7]$:
> $SY := Matrix([R1, R2, R3, R4, R5, R6, R7, R8])$:
> $a := X[0] : b := X[1] : c := X[2] : d := X[3] : e := X[4] : f := X[5]$:
> $g := X[6] : h := X[7]$:
> $VARMAT2 := Matrix([R1, R2, R3, R4, R5, R6, R7, R8])$:
> $a := B[0] : b := B[1] : c := B[2] : d := B[3] : e := B[4] : f := B[5]$:
> $g := B[6] : h := B[7]$:
> $VARMAT3 := Matrix([R1, R2, R3, R4, R5, R6, R7, R8])$:
> $a := C[0] : b := C[1] : c := C[2] : d := C[3] : e := C[4] : f := C[5]$:
> $g := C[6] : h := C[7]$:
> $VARMAT4 := Matrix([R1, R2, R3, R4, R5, R6, R7, R8])$:
>
> **for** row **from** 1 **to** $SIZE$ **do**
> **for** col **from** 1 **to** $SIZE$ **do**
> $ANDY[row, col] := P[row, col] \cdot AN[row, col]$
> $SYD[row, col] := P[row, col] \cdot SY[row, col]$
> $PARAM[row, col] := P[row, col]$:
> **end do**: **end do**:
> **for** xx **from** 1 **to** $SIZE$ **do**
> $P[1, xx] := 1$:
> $P[xx, xx] := 1$:
> **end do**:
>
> $P[7, 2] := \dfrac{P[8, 1]}{P[2, 7]}$: $\quad P[6, 2] := \dfrac{P[7, 1]}{P[2, 6]}$: $\quad P[5, 2] := \dfrac{P[6, 1]}{P[2, 5]}$: $\quad P[4, 2] := \dfrac{P[3, 1]}{P[2, 4]}$:

> $P[3, 2] := \dfrac{P[4, 1]}{P[2, 3]}$: $\quad P[8, 4] := \dfrac{P[5, 1]}{P[4, 8]}$: $\quad P[4, 3] := \dfrac{P[2, 1]}{P[3, 4]}$: $\quad P[8, 5] := \dfrac{P[2, 1]}{P[5, 8]}$:

> $P[6, 5] := \dfrac{P[2, 1]}{P[5, 6]}$: $\quad P[7, 6] := \dfrac{P[2, 1]}{P[6, 7]}$: $\quad P[8, 7] := \dfrac{P[2, 1]}{P[7, 8]}$: $\quad P[7, 5] := \dfrac{P[3, 1]}{P[5, 7]}$:

> $P[8, 6] := \dfrac{P[3, 1]}{P[6, 8]}$: $\quad P[7, 3] := \dfrac{P[5, 1]}{P[3, 7]}$: $\quad P[5, 3] := \dfrac{P[5, 1]}{P[3, 5]}$: $\quad P[6, 4] := \dfrac{P[5, 1]}{P[4, 6]}$:

> $P[8, 2] := \dfrac{P[5, 1]}{P[2, 8]}$: $\quad P[8, 3] := \dfrac{P[6, 1]}{P[3, 8]}$: $\quad P[6, 3] := \dfrac{P[6, 1]}{P[3, 6]}$: $\quad P[7, 4] := \dfrac{P[6, 1]}{P[4, 7]}$:

> $P[5, 4] := \dfrac{P[6, 1]}{P[4, 5]}$: $P[4, 1] := P[2, 1] : P[7, 1] := P[5, 1] : P[8, 1] := P[6, 1]$:

> $P[7, 8] := \dfrac{P[4, 5] \cdot P[5, 1]}{P[6, 1]}$: $\quad P[6, 8] := \dfrac{P[2, 4] \cdot P[4, 5] \cdot P[5, 1] \cdot P[6, 7]}{P[2, 1] \cdot P[6, 1]}$:

> $P[5, 8] := \dfrac{P[3, 4] \cdot P[4, 5] \cdot P[5, 1] \cdot P[5, 7]}{P[3, 1] \cdot P[6, 1]}$: $\quad P[4, 8] := \dfrac{P[4, 5] \cdot P[4, 7] \cdot P[5, 1]^2}{P[4, 6] \cdot P[6, 1]^2}$:

> $P[3, 8] := \dfrac{P[3, 7] \cdot P[4, 5] \cdot P[5, 1]}{P[4, 7] \cdot P[5, 1]}$: $\quad P[2, 8] := \dfrac{P[2, 7] \cdot P[4, 6] \cdot P[6, 1]}{P[4, 7] \cdot P[6, 1]}$:

> $P[3, 4] := P[2, 4] : P[6, 7] := \dfrac{P[2, 1] \cdot P[5, 7]}{P[2, 4] \cdot P[5, 6]}$:

$$> P[5,7] := \frac{P[2,4] \cdot P[4,5] \cdot P[4,7] \cdot P[5,1] \cdot P[5,6]}{P[2,1] \cdot P[4,6] \cdot P[6,1]} \ : \ P[4,6] := \frac{P[2,6] \cdot P[4,5]}{P[2,7]} \ :$$

$$> P[4,5] := \frac{P[2,1]}{P[2,6]} \ : \ P[3,7] := \frac{P[2,6] \cdot P[2,7] \cdot P[3,1]}{P[2,1] \cdot P[2,4]} \ : \ P[5,6] := \frac{P[4,7] \cdot P[5,1]}{P[6,1]} \ :$$

$$> P[2,4] := \frac{P[2,3] \cdot P[3,1]}{P[2,1]} \ : \ P[4,7] := \frac{P[2,1] \cdot P[2,7] \cdot P[3,6]}{P[2,3] \cdot P[3,1]} \ :$$

$$> P[2,7] := \frac{P[2,3] \cdot P[3,1]}{P[2,6] \cdot P[3,5]} \ : \ P[3,6] := \frac{P[2,5] \cdot P[2,6]}{P[2,3]} \ :$$

$$> P[3,1] := \frac{P[2,6]^2 \cdot P[3,5]^2 \cdot P[6,1]^2}{P[2,5]^2 \cdot P[5,1]^2} \ :$$

The order twelve dicyclic group:

The order twelve dicyclic group has Standard form Cayley table:

$$Q_6 \sim \begin{bmatrix}
a & b & c & d & e & f & g & h & i & j & k & l \\
f & a & b & c & d & e & h & i & j & k & l & g \\
e & f & a & b & c & d & i & j & k & l & g & h \\
d & e & f & a & b & c & j & k & l & g & h & i \\
c & d & e & f & a & b & k & l & g & h & i & j \\
b & c & d & e & e & a & l & g & h & i & j & k \\
j & k & l & g & h & i & a & b & c & d & e & f \\
k & l & g & h & i & j & f & a & b & c & d & e \\
l & g & h & i & j & k & e & f & a & b & c & d \\
g & h & i & j & k & l & d & e & f & a & b & c \\
h & i & j & k & l & g & c & d & e & f & a & b \\
i & j & k & l & g & h & b & c & d & e & f & a
\end{bmatrix} \qquad (21.8)$$

The order twelve dicyclic group has one C_2 subgroup, $\{a,d\}$, one C_3 subgroup, $\{a,c,e\}$, three C_4 subgroups, $\{a,d,g,j\}$, $\{a,d,h,k\}$, $\{a,d,i,l\}$, and one C_6 subgroup, $\{a,b,c,d,e,f\}$.

Despite several attempts, your author is unable to calculate the algebraic matrix form of this group – but see *Stop Press* below. Perhaps the reader will be able to do this calculation. If so, your author would be like to be made aware of the result.

Since this group has a C_4 subgroup (three of them actually), it cannot hold a geometric space.

Stop Press: The algebraic matrix form of the order twelve dicyclic has now been calculated. The order twelve dicyclic group contains 6 non-isomorphic algebras. They are of the forms:

$$256 \quad \textit{off} \quad a + b\sqrt[6]{+1} + c\sqrt[3]{+1} + d\sqrt[2]{+1} + e\sqrt[3]{+1} + f\sqrt[6]{+1} + g\sqrt[4]{+1} + h\sqrt[4]{+1} + i\sqrt[4]{+1} + j\sqrt[4]{+1} + k\sqrt[4]{+1} + l\sqrt[4]{+1}$$

$$(21.9)$$

$$256 \quad \textit{off} \quad a+b\sqrt[6]{-1}+c\sqrt[3]{-1}+d\sqrt[2]{-1}+e\sqrt[3]{-1}+f\sqrt[6]{-1}+g\sqrt[4]{-1}+h\sqrt[4]{-1}+i\sqrt[4]{-1}+j\sqrt[4]{-1}+k\sqrt[4]{-1}+l\sqrt[4]{-1}$$
$$(21.10)$$

$$512 \quad \textit{off} \quad a+b\sqrt[6]{+1}+c\sqrt[3]{+1}+d\sqrt[2]{+1}+e\sqrt[3]{-1}+f\sqrt[6]{+1}+g\sqrt[4]{+1}+h\sqrt[4]{+1}+i\sqrt[4]{+1}+j\sqrt[4]{+1}+k\sqrt[4]{+1}+l\sqrt[4]{+1}$$
$$(21.11)$$

$$512 \quad \textit{off} \quad a+b\sqrt[6]{-1}+c\sqrt[3]{-1}+d\sqrt[2]{-1}+e\sqrt[3]{+1}+f\sqrt[6]{-1}+g\sqrt[4]{-1}+h\sqrt[4]{-1}+i\sqrt[4]{-1}+j\sqrt[4]{-1}+k\sqrt[4]{-1}+l\sqrt[4]{-1}$$
$$(21.12)$$

$$256 \quad \textit{off} \quad a+b\sqrt[6]{+1}+c\sqrt[3]{-1}+d\sqrt[2]{+1}+e\sqrt[3]{-1}+f\sqrt[6]{+1}+g\sqrt[4]{+1}+h\sqrt[4]{+1}+i\sqrt[4]{+1}+j\sqrt[4]{+1}+k\sqrt[4]{+1}+l\sqrt[4]{+1}$$
$$(21.13)$$

$$256 \quad \textit{off} \quad a+b\sqrt[6]{-1}+c\sqrt[3]{+1}+d\sqrt[2]{-1}+e\sqrt[3]{+1}+f\sqrt[6]{-1}+g\sqrt[4]{-1}+h\sqrt[4]{-1}+i\sqrt[4]{-1}+j\sqrt[4]{-1}+k\sqrt[4]{-1}+l\sqrt[4]{-1}$$
$$(21.14)$$

All these algebras are non-commutative.

The computer code for the order twelve dicyclic group:
The computer code for the order twelve dicyclic group is:

```
>  R1 := [a, b, c, d, e, f, g, h, i, j, k, l]:
>  R2 := [f, a, b, c, d, e, h, i, j, k, l, g]:
>  R3 := [e, f, a, b, c, d, i, j, k, l, g, h]:
>  R4 := [d, e, f, a, b, c, j, k, l, g, h, i]:
>  R5 := [c, d, e, f, a, b, k, l, g, h, i, j]:
>  R6 := [b, c, d, e, f, a, l, g, h, i, j, k]:
>  R7 := [j, k, l, g, h, i, a, b, c, d, e, f]:
>  R8 := [k, l, g, h, i, j, f, a, b, c, d, e]:
>  R9 := [l, g, h, i, j, k, e, f, a, b, c, d]:
>  R10 := [g, h, i, j, k, l, d, e, f, a, b, c]:
>  R11 := [h, i, j, k, l, g, c, d, e, f, a, b]:
>  R12 := [i, j, k, l, g, h, b, c, d, e, f, a]:
>
>  a := A[0]:  b := A[1]:  c := A[2]:  d := A[3]: e := A[4] :f := A[5]:
>  g := A[6]: h := A[7]: i := A[8] :j := A[9] :k := A[10]: l := A[11]:
>  AN := Matrix([R1, R2, R3, R4, R5, R6, R7, R8, R9, R10, R11, R12]):
>  VARMAT := copy(AN):
>  a := S[0] :b := S[1]: c := S[2]: d := S[3]:  e := S[4]:  f := S[5]:
>   g := S[6]:  h := S[7]:  i := S[8] :j := S[9] :k := S[10]: l := S[11]:
>  SY := Matrix([R1, R2, R3, R4, R5, R6, R7, R8, R9, R10, R11, R12]):
>
>  a := X[0] :b := X[1]: c := X[2]: d := X[3]:  e := X[4]:  f := X[5]:
>   g := X[6]:  h := X[7] :i := X[8] :j := X[9] :k := X[10]: l := X[11]:
>  VARMAT2 := Matrix([R1, R2, R3, R4, R5, R6, R7, R8, R9, R10, R11, R12]):
>  a := B[0] :b := B[1]: c := B[2]: d := B[3]:  e := B[4]:  f := B[5]:
>   g := B[6]:  h := B[7]:  i := B[8] :j := B[9] :k := B[10]: l := B[11]:
>  VARMAT3 := Matrix([R1, R2, R3, R4, R5, R6, R7, R8, R9, R10, R11, R12]):
```

```
>  a := C[0] : b := C[1] : c := C[2] : d := C[3] : e := C[4] : f := C[5] :
>  g := C[6] : h := C[7] : i := C[8] : j := C[9] : k := C[10] : l := C[11] :
>  VARMAT4 := Matrix([R1, R2, R3, R4, R5, R6, R7, R8, R9, R10, R11, R12]) :
>
>  for row from 1 to SIZE do
>  for col from 1 to SIZE do
>  ANDY[row, col] := P[row, col]·AN[row, col]
>  SYD[row, col] := P[row, col]·SY[row, col]
>  PARAM[row, col] := P[row, col] :
>  end do: end do:
>  for xx from 1 to SIZE do
>  P[1, xx] := 1 :
>  P[xx, xx] := 1 :
>  end do:
>
```

Eliminate Parameters:

$$>\ P[6,1] := P[2,1] : \ P[5,1] := P[3,1] : \ P[10,1] := P[7,1] :$$

$$>\ P[11,1] := P[8,1] : \ P[12,1] := P[9,1] :$$

$$>\ P[3,2] := \frac{P[2,1]}{P[2,3]} : \ P[4,3] := \frac{P[2,1]}{P[3,4]} : \ P[5,4] := \frac{P[2,1]}{P[4,5]} : \ P[6,5] := \frac{P[2,1]}{P[5,6]} :$$

$$>\ P[8,7] := \frac{P[2,1]}{P[7,8]} : \ P[9,8] := \frac{P[2,1]}{P[8,9]} : \ P[10,9] := \frac{P[2,1]}{P[9,10]} :$$

$$>\ P[11,10] := \frac{P[2,1]}{P[10,11]} : \ P[12,11] := \frac{P[2,1]}{P[11,12]} : \ P[4,2] := \frac{P[3,1]}{P[2,4]} :$$

$$>\ P[5,3] := \frac{P[3,1]}{P[3,5]} : P[6,4] := \frac{P[3,1]}{P[4,6]} : \ P[9,7] := \frac{P[3,1]}{P[7,9]} : \ P[10,8] := \frac{P[3,1]}{P[8,10]} :$$

$$>\ P[11,9] := \frac{P[3,1]}{P[9,11]} : \ P[12,10] := \frac{P[3,1]}{P[10,12]} : P[5,2] := \frac{P[4,1]}{P[2,5]} :$$

$$>\ P[6,3] := \frac{P[4,1]}{P[3,6]} : \ P[10,7] := \frac{P[4,1]}{P[7,10]} : \ P[11,8] := \frac{P[4,1]}{P[8,11]} :$$

$$>\ P[12,9] := \frac{P[4,1]}{P[9,12]} : P[12,7] := \frac{P[2,1]}{P[7,12]} : \ P[6,2] := \frac{P[3,1]}{P[2,6]} :$$

$$>\ P[11,7] := \frac{P[3,1]}{P[7,11]} : P[12,8] := \frac{P[3,1]}{P[8,12]} :$$

$$>\ P[12,2] := \frac{P[7,1]}{P[2,12]} : \ P[11,3] := \frac{P[7,1]}{P[3,11]} : P[10,4] := \frac{P[7,1]}{P[4,10]} :$$

$$>\ P[9,5] := \frac{P[7,1]}{P[5,9]} : \ P[8,6] := \frac{P[7,1]}{P[6,8]} : \ P[7,4] := \frac{P[7,1]}{P[4,7]} : P[8,3] := \frac{P[7,1]}{P[3,8]} :$$

$$>\ P[9,2] := \frac{P[7,1]}{P[2,9]} : P[11,6] := \frac{P[7,1]}{P[6,11]} : P[12,5] := \frac{P[7,1]}{P[5,12]} :$$

$$>\ P[7,2] := \frac{P[8,1]}{P[2,7]} : \ P[12,3] := \frac{P[8,1]}{P[3,12]} : \ P[11,4] := \frac{P[8,1]}{P[4,11]} :$$

$$>\ P[10,5] := \frac{P[8,1]}{P[5,10]} : P[9,6] := \frac{P[8,1]}{P[6,9]} : P[7,5] := \frac{P[8,1]}{P[5,7]} : P[8,4] := \frac{P[8,1]}{P[4,8]} :$$

$$>\ P[9,3] := \frac{P[8,1]}{P[3,9]} : \ P[10,2] := \frac{P[8,1]}{P[2,10]} : \ P[12,6] := \frac{P[8,1]}{P[6,12]} :$$

$$>\ P[8,2] := \frac{P[9,1]}{P[2,8]} : \ P[7,3] := \frac{P[9,1]}{P[3,7]} : P[12,4] := \frac{P[9,1]}{P[4,12]} : P[11,5] := \frac{P[9,1]}{P[5,11]} :$$

$$>\ P[10,6] := \frac{P[9,1]}{P[6,10]} : \ P[7,6] := \frac{P[9,1]}{P[6,7]} : \ P[8,5] := \frac{P[9,1]}{P[5,8]} : P[9,4] := \frac{P[9,1]}{P[4,9]} :$$

> $P[10,3] := \dfrac{P[9,1]}{P[3,10]} : \quad P[11,2] := \dfrac{P[9,1]}{P[2,11]} : \quad$ # *Leading diagonal complete*

>

> $P[11,12] := \dfrac{P[2,8] \cdot P[8,1]}{P[9,1]} : \quad P[10,12] := \dfrac{P[3,7] \cdot P[7,1]}{P[9,1]} : \quad P[9,12] := P[4,12] :$

> $P[8,12] := \dfrac{P[4,12] \cdot P[8,9]}{P[2,5]} : \quad P[7,12] := \dfrac{P[4,12] \cdot P[7,9]}{P[3,6]} : \quad P[6,12] := \dfrac{P[2,1]}{P[2,11]} :$

> $P[5,12] := \dfrac{P[3,1]}{P[3,10]} : \quad P[4,12] := \dfrac{P[4,1]}{P[4,9]} : \quad P[3,12] := \dfrac{P[3,1]}{P[5,8]} : \quad P[2,12]$

> $\qquad := \dfrac{P[2,1]}{P[6,7]} :$

> $P[10,11] := \dfrac{P[2,7] \cdot P[7,1]}{P[8,1]} : \quad P[9,11] := \dfrac{P[3,5] \cdot P[7,11]}{P[7,9]} : \quad P[8,11] := P[4,11] :$

> $P[7,11] := \dfrac{P[4,11] \cdot P[7,8]}{P[2,5]} : \quad P[6,11] := \dfrac{P[2,1]}{P[2,10]} : \quad P[5,11] := \dfrac{P[3,1]}{P[3,9]} :$

> $P[4,11] := \dfrac{P[4,1]}{P[4,8]} : \quad P[3,11] := \dfrac{P[3,1]}{P[5,7]} : \quad P[2,11] := \dfrac{P[2,8] \cdot P[4,9]}{P[4,8]} :$

> $P[9,10] := \dfrac{P[2,3] \cdot P[8,10]}{P[8,9]} : \quad P[8,10] := \dfrac{P[2,4] \cdot P[7,10]}{P[7,8]} : \quad P[7,10] := P[4,10] :$

> $P[6,10] := \dfrac{P[2,1]}{P[2,9]} : \quad P[5,10] := \dfrac{P[4,10] \cdot P[5,7]}{P[4,8]} : \quad P[4,10] := \dfrac{P[4,1]}{P[4,7]} :$

> $P[3,10] := \dfrac{P[3,7] \cdot P[4,9]}{P[4,7]} : \quad P[2,10] := \dfrac{P[2,7] \cdot P[4,8]}{P[4,7]} : \quad P[8,9] := \dfrac{P[2,3] \cdot P[7,9]}{P[7,8]} :$

> $P[6,9] := \dfrac{P[2,1]}{P[2,8]} : \quad P[5,9] := \dfrac{P[3,1]}{P[3,7]} : \quad P[3,9] := \dfrac{P[3,4] \cdot P[4,8]}{P[2,8]} :$

> $P[2,9] := \dfrac{P[2,4] \cdot P[4,7]}{P[3,7]} : \quad P[3,4] := P[2,4] : \quad P[3,6] := \dfrac{P[2,4] \cdot P[4,1]}{P[2,1]} :$

> $P[6,8] := \dfrac{P[2,1]}{P[2,7]} : \quad P[5,8] := \dfrac{P[2,8] \cdot P[4,9]}{P[2,5]} : \quad P[4,5] := P[2,5] :$

> $P[3,8] := \dfrac{P[2,4] \cdot P[4,7]}{P[2,7]} : \quad P[2,8] := \dfrac{P[2,3] \cdot P[3,7]}{P[2,7]} : \quad P[2,6] := \dfrac{P[2,3] \cdot P[3,1]}{P[2,1]} :$

> $P[2,5] := \dfrac{P[2,4] \cdot P[4,1]}{P[3,1]} : \quad P[6,7] := \dfrac{P[3,7] \cdot P[4,9]}{P[4,6]} : \quad P[4,6] := \dfrac{P[2,4] \cdot P[4,1]}{P[2,1]} :$

> $P[5,6] := \dfrac{P[2,3] \cdot P[3,1]}{P[2,1]} : \quad P[7,9] := \dfrac{P[3,7] \cdot P[4,9] \cdot P[7,1]}{P[4,7] \cdot P[9,1]} :$

> $P[7,8] := \dfrac{P[2,7] \cdot P[4,8] \cdot P[7,1]}{P[4,7] \cdot P[8,1]} : \quad P[5,7] := \dfrac{P[2,7] \cdot P[3,1] \cdot P[4,8]}{P[2,4] \cdot P[4,1]} :$

> $P[3,5] := \dfrac{P[2,4]^2 \cdot P[4,1]}{P[2,3] \cdot P[3,1]} : \quad P[4,9] := \dfrac{P[2,4]^2 \cdot P[4,1] \cdot P[4,7]^2 \cdot P[8,1] \cdot P[9,1]}{P[2,7]^2 \cdot P[3,7]^2 \cdot P[4,8] \cdot P[7,1]^2} :$

> $P[4,1] := \dfrac{P[2,7]^6 \cdot P[4,8]^3 \cdot P[7,1]^3}{P[2,3]^2 \cdot P[2,4]^2 \cdot P[4,7]^3 \cdot P[8,1]^3} :$

The Dihedral Groups

There is a dihedral group of every even order greater than four, $Ord = 2n \; : \; n \geq 3$. One may think of the dihedral groups as a cyclic group of order n which can by reflected - think flat triangle or square or pentagon… and then flip the triangle or square or pentagon … over.

Any group of order $2p$ where p is a prime number is either cyclic or dihedral.

The order eight dihedral group:

The order eight dihedral group, D_4, is a non-commutative group. The group D_4 has one C_4 subgroup and five C_2 subgroups. One of the C_2 subgroups is within the C_4 subgroup.

The Standard form Cayley table of the order eight dihedral group is:

$$D_4 \sim \begin{bmatrix} a & b & c & d & e & f & g & h \\ d & a & b & c & f & g & h & e \\ c & d & a & b & g & h & e & f \\ b & c & d & a & h & e & f & g \\ e & f & g & h & a & b & c & d \\ f & g & h & e & d & a & b & c \\ g & h & e & f & c & d & a & b \\ h & e & f & g & b & c & d & a \end{bmatrix} \tag{22.1}$$

We do not produce the algebraic matrix form because it is too large to fit on the page.

All the parameter elimination equations are linear (only one solution). There are 128 separate algebras, in three non-isomorphic types. That algebra is of the form:

$$a + b\sqrt[4]{\pm 1} + c\sqrt[2]{\pm 1} + d\sqrt[4]{\pm 1} + e\sqrt[2]{\pm 1} + f\sqrt[2]{\pm 1} + g\sqrt[2]{\pm 1} + h\sqrt[2]{\pm 1} \tag{22.2}$$

Which, of course, reflects the subgroups of the quaternion group. The subgroups are:

$$\{a,b,c,d\}, \; \{a,c\}, \; \{a,e\}, \; \{a,f\}, \; \{a,g\}, \; \{a,h\} \tag{22.3}$$

The three types of algebra are:

$$32 \quad \text{off} \quad 1 + 2\sqrt[4]{+1} + 5\sqrt[2]{+1} \tag{22.4}$$

$$64 \quad \text{off} \quad 1 + 2\sqrt[4]{+1} + 3\sqrt[2]{+1} + 2\sqrt[2]{-1} \tag{22.5}$$

$$32 \quad \text{off} \quad 1 + 2\sqrt[4]{+1} + \sqrt[2]{+1} + 4\sqrt[2]{-1} \tag{22.6}$$

The algebras are all non-commutative.

Computer calculations show that there are no emergent distance functions which can support a geometric space. Since this group contains a C_4 subgroup, we know that anyway.

The code of the order eight dihedral group:

The Maple 17 code used to calculate the algebraic matrix form of the D_4 group is:

```
>  R1 := [a, b, c, d, e, f, g, h] :
>  R2 := [d, a, b, c, f, g, h, e] :
>  R3 := [c, d, a, b, g, h, e, f] :
>  R4 := [b, c, d, a, h, e, f, g] :
>  R5 := [e, f, g, h, a, b, c, d] :
>  R6 := [f, g, h, e, d, a, b, c] :
>  R7 := [g, h, e, f, c, d, a, b] :
>  R8 := [h, e, f, g, b, c, d, a] :
>
>  a := A[0] :  b := A[1] : c := A[2] : d := A[3] : e := A[4] : f := A[5] :
>  g := A[6] : h := A[7] :
>  AN := Matrix([R1, R2, R3, R4, R5, R6, R7, R8]) :
>  VARMAT := copy(AN) :
>  a := S[0] : b := S[1] : c := S[2] : d := S[3] : e := S[4] : f := S[5] :
>  g := S[6] : h := S[7] :
>  SY := Matrix([R1, R2, R3, R4, R5, R6, R7, R8]) :
>  a := X[0] : b := X[1] : c := X[2] : d := X[3] : e := X[4] : f := X[5] :
>  g := X[6] : h := X[7] :
>  VARMAT2 := Matrix([R1, R2, R3, R4, R5, R6, R7, R8]) :
>  a := B[0] : b := B[1] : c := B[2] : d := B[3] : e := B[4] : f := B[5] :
>  g := B[6] : h := B[7] :
>  VARMAT3 := Matrix([R1, R2, R3, R4, R5, R6, R7, R8]) :
>  a := C[0] : b := C[1] : c := C[2] : d := C[3] : e := C[4] : f := C[5] :
>  g := C[6] : h := C[7] :
>  VARMAT4 := Matrix([R1, R2, R3, R4, R5, R6, R7, R8]) :
>
>  for row from 1 to SIZE do
>  for col from 1 to SIZE do
>  ANDY[row, col] := P[row, col]·AN[row, col]
>  SYD[row, col] := P[row, col]·SY[row, col]
>  PARAM[row, col] := P[row, col] :
>  end do: end do:
>  for xx from 1 to SIZE do
>  P[1, xx] := 1 :
>  P[xx, xx] := 1 :
>  end do:
>  VARMAT;
```

$$> \quad P[3, 2] := \frac{P[4, 1]}{P[2, 3]} : \quad P[4, 2] := \frac{P[3, 1]}{P[2, 4]} : \quad P[5, 2] := \frac{P[6, 1]}{P[2, 5]} :$$

> $P[6,2] := \dfrac{P[7,1]}{P[2,6]}$: $P[7,2] := \dfrac{P[8,1]}{P[2,7]}$: $P[8,2] := \dfrac{P[5,1]}{P[2,8]}$:

> $P[6,4] := \dfrac{P[5,1]}{P[4,6]}$: $P[4,3] := \dfrac{P[2,1]}{P[3,4]}$: $P[7,3] := \dfrac{P[5,1]}{P[3,7]}$:

> $P[6,5] := \dfrac{P[2,1]}{P[5,6]}$: $P[7,5] := \dfrac{P[3,1]}{P[5,7]}$: $P[8,5] := \dfrac{P[4,1]}{P[5,8]}$:

> $P[7,6] := \dfrac{P[2,1]}{P[6,7]}$: $P[8,6] := \dfrac{P[3,1]}{P[6,8]}$: $P[8,7] := \dfrac{P[2,1]}{P[7,8]}$:

> $P[4,1] := P[2,1]$: $P[8,3] := \dfrac{P[6,1]}{P[3,8]}$: $P[5,3] := \dfrac{P[7,1]}{P[3,5]}$:

> $P[6,3] := \dfrac{P[8,1]}{P[3,6]}$: $P[7,4] := \dfrac{P[6,1]}{P[4,7]}$: $P[8,4] := \dfrac{P[7,1]}{P[4,8]}$:

> $P[5,4] := \dfrac{P[8,1]}{P[4,5]}$: # *leading dagonal done*

> $P[7,8] := \dfrac{P[2,7]\cdot P[7,1]}{P[8,1]}$: $P[6,8] := \dfrac{P[2,7]\cdot P[3,1]\cdot P[6,7]\cdot P[7,1]}{P[2,1]\cdot P[2,4]\cdot P[8,1]}$: :

> $P[5,8] := \dfrac{P[2,1]\cdot P[2,7]\cdot P[5,7]\cdot P[7,1]}{P[2,3]\cdot P[3,1]\cdot P[8,1]}$: $P[4,8] := \dfrac{P[2,7]\cdot P[4,7]\cdot P[7,1]^2}{P[2,6]\cdot P[6,1]\cdot P[8,1]}$:

> $P[3,8] := \dfrac{P[2,7]\cdot P[3,7]\cdot P[6,1]\cdot P[7,1]}{P[2,5]\cdot P[5,1]\cdot P[8,1]}$: $P[6,7] := \dfrac{P[2,6]\cdot P[6,1]}{P[7,1]}$:

> $P[5,7] := \dfrac{P[2,6]\cdot P[3,1]\cdot P[5,6]\cdot P[6,1]}{P[2,1]\cdot P[2,4]\cdot P[7,1]}$: $P[4,7] := \dfrac{P[2,6]\cdot P[4,6]\cdot P[6,1]^2}{P[2,5]\cdot P[5,1]\cdot P[7,1]}$:

> $P[3,7] := \dfrac{P[2,6]\cdot P[3,6]\cdot P[5,1]\cdot P[6,1]}{P[2,8]\cdot P[7,1]\cdot P[8,1]}$: $P[5,6] := \dfrac{P[2,5]\cdot P[5,1]}{P[6,1]}$:

> $P[4,6] := \dfrac{P[2,5]\cdot P[4,5]\cdot P[5,1]^2}{P[2,8]\cdot P[6,1]\cdot P[8,1]}$: $P[3,6] := \dfrac{P[2,5]\cdot P[3,5]\cdot P[5,1]\cdot P[8,1]}{P[2,7]\cdot P[6,1]\cdot P[7,1]}$:

> $P[4,5] := \dfrac{P[2,1]}{P[2,8]}$: $P[3,5] := \dfrac{P[2,1]\cdot P[2,7]\cdot P[3,4]\cdot P[5,1]\cdot P[7,1]}{P[2,8]^3\cdot P[8,1]^2}$:

> $P[2,8] := \dfrac{P[2,1]\cdot P[2,3]\cdot P[2,4]}{P[2,5]\cdot P[2,6]\cdot P[2,7]}$: $P[3,4] := \dfrac{P[2,3]\cdot P[3,1]}{P[2,1]}$:

> $P[2,4] := \dfrac{P[2,3]\cdot P[3,1]}{P[2,1]}$: $P[7,1] := \dfrac{P[2,6]^2\cdot P[6,1]^2}{P[2,5]^2\cdot P[5,1]}$:

> $P[8,1] := \dfrac{P[2,6]^2\cdot P[2,7]^2 P[6,1]}{P[2,3]^2\cdot P[3,1]}$: $P[3,1] := \dfrac{P[2,5]^4\cdot P[5,1]^2}{P[2,3]^2\cdot P[6,1]^2}$:

The order ten dihedral group:

The order ten dihedral group, D_5, is a non-commutative group. The group D_5 has one C_5 subgroup and five C_2 subgroups.

The Standard form Cayley table of the order ten dihedral group is:

$$
D_5 \sim
\begin{bmatrix}
a & b & c & d & e & f & g & h & i & j \\
e & a & b & c & d & g & h & i & j & f \\
d & e & a & b & c & h & i & j & f & g \\
c & d & e & a & b & i & j & f & g & h \\
b & c & d & e & a & j & f & g & h & i \\
f & g & h & i & j & a & b & c & d & e \\
g & h & i & j & f & e & a & b & c & d \\
h & i & j & f & g & d & e & a & b & c \\
i & j & f & g & h & c & d & e & a & b \\
j & f & g & h & i & b & c & d & e & a
\end{bmatrix}
\qquad (22.7)
$$

This group has five C_2 subgroups and one C_5 subgroup.

Your author has tried several times and failed to calculate the algebraic matrix form of this group. In these calculations, we eventually come to a quadratic parameter elimination equation containing odd powered parameters; this will introduce into the algebraic matrix form the imaginary number $i = \sqrt[2]{-1}$ when the parameters with the odd power are set to minus unity. This is obviously not acceptable.

Geometric spaces and the dihedral groups:

Every dihedral group of order Ord has a cyclic subgroup of order $\dfrac{Ord}{2}$. We therefore know that the dihedral groups $D_3...D_{30}$ do not contain a geometric space because we have examined the cyclic groups up to order fifteen. Further, since we know that all cyclic groups other than cyclic groups of order $2p$ where p is prime cannot hold a geometric space, and so the corresponding dihedral groups cannot hold a geometric space.

At the end of the chapter on cyclic groups, we found that no cyclic group of order greater than two can hold a geometric space. Hence, no dihedral group can hold a geometric space.

The Order Twelve Alternating Group

Your author has been unable to satisfactorily discover the algebraic matrix form of the A_4 finite group. However, he has found an 'unsatisfactory' algebraic matrix form. By 'unsatisfactory', your author means that this 'unsatisfactory' algebraic matrix form of the order twelve A_4 group has only ten parameters rather than the expected eleven. The reader is reminded that we have no proof of the conjecture that every finite group holds more division algebras than the basic 'all parameters are plus one' division algebra or of the conjecture that every finite group holds an algebraic matrix form with $(n-1)$ parameters thereby holding at least 2^{n-1} separate division algebras. However, given the evidence of these conjectures within the lowest order finite groups, we 'feel' that these conjectures will be correct. It is thus very intriguing that your author has found an A_4 algebraic matrix form with only ten parameters. We might expect that by commenting out in turn each one of the parameter elimination equations, we would be able to find an eleven parameter algebraic matrix form. Your author has tried this; it does not work.

It is your author's opinion that he has made an error somewhere in this calculation of the A_4 algebraic matrix form, but, in spite of thorough checking, he can find no error. This reflects the current level of understanding of this new area of mathematics. There is much opportunity for young mathematicians to develop this area of mathematics.

In spite of your author's opinion that he has made an error, he feels it proper to report what he has found to the reader, and so this chapter does make such report.

The A_4 group:

The alternating groups are the groups of even permutations. The group A_n is always a normal subgroup of the symmetric group of all permutations S_n. The orders of the alternating groups, A_n, are $\frac{n!}{2}$. The group A_4 is exceptional within the alternating groups in that all alternating groups are finite simple groups that have no normal proper subgroups whereas A_4 has the group $C_2 \times C_2$ as a normal subgroup.

Lagrange's theorem states that the order of a subgroup must be a divisor of the order of the group. The converse of this is not true; it is not true that a group must have a subgroup of order equal to every divisor of the order of the group. The group A_4 is the lowest order group that lacks a subgroup corresponding to every divisor of the order of the group in that A_4, which is of order twelve, has no order six subgroup. It is true that if p is a prime divisor of the order of a group, then that group does have a subgroup of order p. The proper subgroups of the A_4 group are two C_2 subgroups, four C_3 subgroups and one $C_2 \times C_2$ subgroup. The A_4 group is a non-commutative group.

The Standard form Cayley table of A_4 is:

$$
A_4 \sim
\begin{bmatrix}
a & b & c & d & e & f & g & h & i & j & k & l \\
b & a & d & c & h & g & f & e & k & l & i & j \\
c & d & a & b & f & e & h & g & l & k & j & i \\
d & c & b & a & g & h & e & f & j & i & l & k \\
i & j & k & l & a & b & c & d & e & f & g & h \\
k & l & i & j & b & a & d & c & h & g & f & e \\
l & k & j & i & c & d & a & b & f & e & h & g \\
j & i & l & k & d & c & b & a & g & h & e & f \\
e & f & g & h & i & j & k & l & a & b & c & d \\
h & g & f & e & k & l & i & j & b & a & d & c \\
f & e & h & g & l & k & j & i & c & d & a & b \\
g & h & e & f & j & i & l & k & d & c & b & a
\end{bmatrix}
\qquad (23.1)
$$

The presence of the $C_2 \times C_2$ subgroup appears as three copies of the $C_2 \times C_2$ Standard form Cayley table as three 4×4 blocks along the leading diagonal. The reader will recall that the $C_2 \times C_2$ group holds the only two geometric spaces we have discovered so far.

The A_4 algebraic matrix form:

We have a multiplicatively closed algebraic matrix form for the A_4 group, but we wonder if it is the optimal algebraic matrix form. Since the A_4 group is of order twelve, we might expect that we would have eleven parameters left when we reach multiplicative closure. With this algebraic matrix form, we have only ten parameters left when we reach multiplicative closure. Is this 'absence of a parameter' some human error somewhere? If we comment out each individual elimination equation one at a time, we find that we can dispense with none of the elimination equations if we are to get multiplicative closure. Since the parameters scale the imaginary axes against the real axis, we think we must have made an error, but we have checked carefully and can find no error.

If we alter existing successful programs which have the expected number of parameters by eliminating an extra parameter, we find that the program still works and still produces the algebras but in different numbers. With one less parameter, we will have only half the number of algebras, of course. Such an extra elimination equation can be commented out without losing multiplicative closure. In the extreme, putting all parameters equal to $P_{2,1}$ and putting $P_{2,1} = 1$ will produce, by taking the exponential, the 'all parameters are plus unity' division algebra.

What we are missing, is an understanding of parameter elimination and with it a systematic way to eliminate parameters.

A quadratic parameter elimination equation:

The final parameter elimination equation of the process of deriving this ten parameter A_4 algebraic matrix form is a quadratic equation giving plus and minus results. Both results are multiplicatively closed, and so we get the same total number of algebras as we would have with eleven remaining parameters and no quadratic elimination equation.

Your author's computer is unable to deal with all the possible algebras within this ten parameter algebraic matrix form before next Christmas, and so we stop the program before it finished to get a taste of the algebras this group contains. The list we have so far is probably not all the algebras of the A_4 group but includes:

With the quadratic parameter taking the negative root:

$$a + 3\sqrt[2]{-1} + 4\sqrt[3]{+1} + 4\sqrt[3]{-1} \qquad Non-Commutative \qquad (23.2)$$

$$a + 3\sqrt[2]{-1} + 5\sqrt[3]{+1} + 3\sqrt[3]{-1} \qquad Non-Commutative \qquad (23.3)$$

$$a + 3\sqrt[2]{-1} + 6\sqrt[3]{+1} + 2\sqrt[3]{-1} \qquad Non-Commutative \qquad (23.4)$$

$$a + 3\sqrt[2]{-1} + 3\sqrt[3]{+1} + 5\sqrt[3]{-1} \qquad Non-Commutative \qquad (23.5)$$

$$a + 3\sqrt[2]{-1} + 2\sqrt[3]{+1} + 6\sqrt[3]{-1} \qquad Non-Commutative \qquad (23.6)$$

$$a + 3\sqrt[2]{-1} + \sqrt[3]{+1} + 7\sqrt[3]{-1} \qquad Non-Commutative \qquad (23.7)$$

$$a + 3\sqrt[2]{-1} + 7\sqrt[3]{+1} + \sqrt[3]{-1} \qquad Non-Commutative \qquad (23.8)$$

$$a + 3\sqrt[2]{-1} + 8\sqrt[3]{+1} \qquad Non-Commutative \qquad (23.9)$$

$$a + 3\sqrt[2]{-1} + 8\sqrt[3]{-1} \qquad Non-Commutative \qquad (23.10)$$

With the quadratic parameter taking the positive root:

$$a + 3\sqrt[2]{+1} + 4\sqrt[3]{+1} + 4\sqrt[3]{-1} \qquad Non-Commutative \qquad (23.11)$$

$$a + 3\sqrt[2]{+1} + 5\sqrt[3]{+1} + 3\sqrt[3]{-1} \qquad Non-Commutative \qquad (23.12)$$

$$a + 3\sqrt[2]{+1} + 6\sqrt[3]{+1} + 2\sqrt[3]{-1} \qquad Non-Commutative \qquad (23.13)$$

$$a + 3\sqrt[2]{+1} + 3\sqrt[3]{+1} + 5\sqrt[3]{-1} \qquad Non-Commutative \qquad (23.14)$$

$$a + 3\sqrt[2]{+1} + 2\sqrt[3]{+1} + 6\sqrt[3]{-1} \qquad Non-Commutative \qquad (23.15)$$

$$a + 3\sqrt[2]{+1} + \sqrt[3]{+1} + 7\sqrt[3]{-1} \qquad Non-Commutative \qquad (23.16)$$

$$a + 3\sqrt[2]{+1} + 7\sqrt[3]{+1} + \sqrt[3]{-1} \qquad Non-Commutative \qquad (23.17)$$

$$a + 3\sqrt[2]{+1} + 8\sqrt[3]{+1} \qquad Non-Commutative \qquad (23.18)$$

$$a + 3\sqrt[2]{+1} + 8\sqrt[3]{-1} \qquad Non-Commutative \qquad (23.19)$$

Since the A_4 group has C_3 subgroups, we expect no geometric spaces within this group.

The computer code:

The computer code which leaves only ten parameters is:

```
> R1 := [a, b, c, d, e, f, g, h, i, j, k, l] :
> R2 := [b, a, d, c, h, g, f, e, k, l, i, j] :
> R3 := [c, d, a, b, f, e, h, g, l, k, j, i] :
> R4 := [d, c, b, a, g, h, e, f, j, i, l, k] :
> R5 := [i, j, k, l, a, b, c, d, e, f, g, h] :
> R6 := [k, l, i, j, b, a, d, c, h, g, f, e] :
> R7 := [l, k, j, i, c, d, a, b, f, e, h, g] :
> R8 := [j, i, l, k, d, c, b, a, g, h, e, f] :
> R9 := [e, f, g, h, i, j, k, l, a, b, c, d] :
> R10 := [h, g, f, e, k, l, i, j, b, a, d, c] :
> R11 := [f, e, h, g, l, k, j, i, c, d, a, b] :
> R12 := [g, h, e, f, j, i, l, k, d, c, b, a] :
>
> a := A[0] : b := A[1] : c := A[2] : d := A[3] : e := A[4] : f := A[5] :
> g := A[6] : h := A[7] : i := A[8] : j := A[9] : k := A[10] : l := A[11] :
> AN := Matrix([R1, R2, R3, R4, R5, R6, R7, R8, R9, R10, R11, R12]) :
> VARMAT := copy(AN) :
> a := S[0] : b := S[1] : c := S[2] : d := S[3] : e := S[4] : f := S[5] :
> g := S[6] : h := S[7] : i := S[8] : j := S[9] : k := S[10] : l := S[11] :
> SY := Matrix([R1, R2, R3, R4, R5, R6, R7, R8, R9, R10, R11, R12]) :
> a := X[0] : b := X[1] : c := X[2] : d := X[3] : e := X[4] : f := X[5] :
> g := X[6] : h := X[7] : i := X[8] : j := X[9] : k := X[10] : l := X[11] :
> VARMAT2 := Matrix([R1, R2, R3, R4, R5, R6, R7, R8, R9, R10, R11, R12]) :
> a := B[0] : b := B[1] : c := B[2] : d := B[3] : e := B[4] : f := B[5] :
> g := B[6] : h := B[7] : i := B[8] : j := B[9] : k := B[10] : l := B[11] :
> VARMAT3 := Matrix([R1, R2, R3, R4, R5, R6, R7, R8, R9, R10, R11, R12]) :
> a := C[0] : b := C[1] : c := C[2] : d := C[3] : e := C[4] : f := C[5] :
> g := C[6] : h := C[7] : i := C[8] : j := C[9] : k := C[10] : l := C[11] :
> VARMAT4 := Matrix([R1, R2, R3, R4, R5, R6, R7, R8, R9, R10, R11, R12]) :
>
> for row from 1 to SIZE do
> for col from 1 to SIZE do
> ANDY[row, col] := P[row, col]·AN[row, col]
> SYD[row, col] := P[row, col]·SY[row, col]
> PARAM[row, col] := P[row, col] :
> end do: end do:
> for xx from 1 to SIZE do
> P[1, xx] := 1 :
> P[xx, xx] := 1 :
> end do:
>
```

$$> \quad P[4, 2] := \frac{P[3, 1]}{P[2, 4]} : \quad P[3, 2] := \frac{P[4, 1]}{P[2, 3]} : \quad P[4, 3] := \frac{P[2, 1]}{P[3, 4]} :$$

> $P[6,5] := \dfrac{P[2,1]}{P[5,6]}$: $P[8,7] := \dfrac{P[2,1]}{P[7,8]}$: $P[10,9] := \dfrac{P[2,1]}{P[9,10]}$:

> $P[12,11] := \dfrac{P[2,1]}{P[11,12]}$: $P[12,10] := \dfrac{P[3,1]}{P[10,12]}$: $P[11,9] := \dfrac{P[3,1]}{P[9,11]}$:

> $P[7,5] := \dfrac{P[3,1]}{P[5,7]}$: $P[8,6] := \dfrac{P[3,1]}{P[6,8]}$: $P[8,5] := \dfrac{P[4,1]}{P[5,8]}$:

> $P[7,6] := \dfrac{P[4,1]}{P[6,7]}$: $P[12,9] := \dfrac{P[4,1]}{P[9,12]}$: $P[11,10] := \dfrac{P[4,1]}{P[10,11]}$:

> $P[8,2] := \dfrac{P[5,1]}{P[2,8]}$: $P[7,2] := \dfrac{P[6,1]}{P[2,7]}$: $P[6,2] := \dfrac{P[7,1]}{P[2,6]}$:

> $P[5,2] := \dfrac{P[8,1]}{P[2,5]}$: $P[11,2] := \dfrac{P[9,1]}{P[2,11]}$: $P[12,2] := \dfrac{P[10,1]}{P[2,12]}$:

> $P[10,2] := \dfrac{P[12,1]}{P[2,10]}$: $P[9,2] := \dfrac{P[11,1]}{P[2,9]}$: $P[6,3] := \dfrac{P[5,1]}{P[3,6]}$:

> $P[5,3] := \dfrac{P[6,1]}{P[3,5]}$: $P[8,3] := \dfrac{P[7,1]}{P[3,8]}$: $P[7,3] := \dfrac{P[8,1]}{P[3,7]}$:

> $P[12,3] := \dfrac{P[9,1]}{P[3,12]}$: $P[11,3] := \dfrac{P[10,1]}{P[3,11]}$: $P[9,3] := \dfrac{P[12,1]}{P[3,9]}$:

> $P[10,3] := \dfrac{P[11,1]}{P[3,10]}$: $P[7,4] := \dfrac{P[5,1]}{P[4,7]}$: $P[8,4] := \dfrac{P[6,1]}{P[4,8]}$:

> $P[5,4] := \dfrac{P[7,1]}{P[4,5]}$: $P[6,4] := \dfrac{P[8,1]}{P[4,6]}$: $P[10,4] := \dfrac{P[9,1]}{P[4,10]}$:

> $P[9,4] := \dfrac{P[10,1]}{P[4,9]}$: $P[12,4] := \dfrac{P[11,1]}{P[4,12]}$: $P[11,4] := \dfrac{P[12,1]}{P[4,11]}$:

> $P[9,5] := \dfrac{P[5,1]}{P[5,9]}$: $P[10,5] := \dfrac{P[6,1]}{P[5,10]}$: $P[11,5] := \dfrac{P[7,1]}{P[5,11]}$:

> $P[12,5] := \dfrac{P[8,1]}{P[5,12]}$: $P[9,1] := P[5,1]$: $P[10,1] := P[8,1]$:

> $P[11,1] := P[6,1]$: $P[12,1] := P[7,1]$: $P[9,6] := \dfrac{P[8,1]}{P[6,9]}$:

> $P[12,6] := \dfrac{P[5,1]}{P[6,12]}$: $P[11,6] := \dfrac{P[6,1]}{P[6,11]}$: $P[10,6] := \dfrac{P[7,1]}{P[6,10]}$:

> $P[12,7] := \dfrac{P[7,1]}{P[7,12]}$: $P[11,7] := \dfrac{P[8,1]}{P[7,11]}$: $P[10,7] := \dfrac{P[5,1]}{P[7,10]}$:

> $P[9,7] := \dfrac{P[6,1]}{P[7,9]}$: $P[11,8] := \dfrac{P[5,1]}{P[8,11]}$: $P[10,8] := \dfrac{P[6,1]}{P[8,10]}$:

> $P[12,8] := \dfrac{P[6,1]}{P[8,12]}$: $P[9,8] := \dfrac{P[7,1]}{P[8,9]}$: $P[8,1] := P[6,1]$: # *end of lead diagonal*

>

> $P[11,12] := \dfrac{P[2,6] \cdot P[6,1]}{P[7,1]}$: $P[10,12] := \dfrac{P[3,8] \cdot P[6,1]}{P[7,1]}$:

> $P[9,12] := \dfrac{P[4,5] \cdot P[5,1]}{P[7,1]}$:

> $P[8,12] := \dfrac{P[6,1] \cdot P[6,10]}{P[7,1]}$: $P[7,12] := \dfrac{P[2,12] \cdot P[7,11]}{P[2,7]}$:

> $P[6,12] := \dfrac{P[5,11] \cdot P[6,1]}{P[7,1]}$:

> $P[5,12] := \dfrac{P[5,1] \cdot P[8,9]}{P[7,1]}$: $P[4,12] := \dfrac{P[4,1]}{P[4,11]}$:

> $P[3,12] := \dfrac{P[3,1]}{P[3,9]}$:

> $P[2, 12] := \dfrac{P[2, 1]}{P[2, 10]}$: $P[10, 11] := P[4, 8]$:

> $P[9, 11] := \dfrac{P[3, 5] \cdot P[5, 1]}{P[6, 1]}$: $P[8, 11] := P[5, 10]$: $P[7, 11] := P[6, 10]$:

> $P[6, 11] := \dfrac{P[6, 7] \cdot P[6, 10]}{P[4, 6]}$: $P[5, 11] := \dfrac{P[4, 8] \cdot P[5, 10]}{P[6, 7]}$:

> $P[4, 11] := \dfrac{P[6, 7] \cdot P[7, 1]}{P[6, 1]}$:

> $P[3, 11] := \dfrac{P[3, 1]}{P[3, 10]}$: $P[2, 11] := \dfrac{P[2, 1]}{P[2, 9]}$:

> $P[9, 10] := \dfrac{P[2, 5] \cdot P[5, 1]}{P[6, 1]}$: $P[8, 10] := \dfrac{P[3, 10] \cdot P[6, 10]}{P[3, 8]}$:

> $P[7, 10] := \dfrac{P[5, 1] \cdot P[8, 9]}{P[6, 1]}$:

> $P[6, 10] := \dfrac{P[4, 10] \cdot P[8, 9]}{P[4, 6]}$: $P[5, 10] := \dfrac{P[5, 1] \cdot P[6, 9]}{P[6, 1]}$:

> $P[4, 10] := \dfrac{P[4, 1]}{P[4, 9]}$:

> $P[3, 10] := \dfrac{P[3, 7] \cdot P[8, 9]}{P[6, 9]}$: $P[8, 9] := \dfrac{P[2, 1] \cdot P[7, 9]}{P[2, 7] \cdot P[7, 8]}$:

> $P[7, 9] := \dfrac{P[4, 8] \cdot P[6, 9]}{P[6, 7]}$: $P[6, 9] := \dfrac{P[2, 1] \cdot P[5, 9]}{P[2, 8] \cdot P[5, 6]}$:

> $P[4, 9] := \dfrac{P[2, 1] \cdot P[3, 9]}{P[2, 10] \cdot P[3, 4]}$:

> $P[7, 8] := \dfrac{P[2, 1] \cdot P[7, 1]}{P[2, 10] \cdot P[6, 1]}$: $P[6, 8] := \dfrac{P[2, 1] \cdot P[5, 8]}{P[2, 3] \cdot P[5, 6]}$:

> $P[5, 8] := \dfrac{P[3, 8] \cdot P[6, 7]}{P[3, 5]}$:

> $P[4, 8] := \dfrac{P[4, 1]}{P[4, 6]}$: $P[3, 8] := \dfrac{P[3, 1]}{P[3, 7]}$: $P[2, 8] := \dfrac{P[2, 1]}{P[2, 5]}$:

> $P[3, 4] := \dfrac{P[2, 3] \cdot P[3, 1]}{P[4, 1]}$: $P[4, 5] := \dfrac{P[4, 1]}{P[4, 7]}$:

> $P[3, 5] := \dfrac{P[3, 1]}{P[3, 6]}$:

> $P[5, 6] := \dfrac{P[2, 9] \cdot P[5, 1]}{P[6, 1]}$: $P[4, 6] := \dfrac{P[2, 6] \cdot P[3, 7]}{P[2, 4]}$:

> $P[3, 6] := \dfrac{P[2, 6] \cdot P[4, 7]}{P[2, 3]}$:

> $P[2, 6] := \dfrac{P[2, 1]}{P[2, 7]}$: $P[5, 7] := \dfrac{P[3, 9] \cdot P[5, 1]}{P[7, 1]}$:

> $P[2, 4] := \dfrac{P[2, 1]}{P[2, 3]}$:

> $P[3, 9] := \dfrac{P[2, 9] \cdot P[6, 7] \cdot P[7, 1]}{P[2, 3] \cdot P[6, 1]}$:

> $P[3, 7] := \dfrac{P[2, 3] \cdot P[2, 5] \cdot P[3, 1] \cdot P[4, 7]}{P[2, 1] \cdot P[4, 1]}$:

> $P[2, 10] := \dfrac{P[2, 5] \cdot P[2, 7] \cdot P[2, 9] \cdot P[5, 1] \cdot P[7, 1]}{P[2, 1] \cdot P[6, 1]^2}$:

> $P[4, 1] := \dfrac{P[2, 1] \cdot P[6, 7]^2}{P[2, 7]^2}$:

> $P[3,1] := \dfrac{P[2,1] \cdot P[2,9]^2 \cdot P[5,1]^2 \cdot P[6,7]^4}{P[2,3]^2 \cdot P[2,7]^2 \cdot P[4,7]^2 \cdot P[6,1]^2}$:

>

> $PM2 := 1$:   *# The next, and final, elimination equation is a quadratic.*

> $P[2,1] := (-1)^{PM2} \cdot \dfrac{P[2,9]^2 \cdot P[5,1]^2 \cdot P[6,7]^2}{P[4,7]^2 \cdot P[6,1]^2}$:

>

> *#shift:=2:*

> *#JACK:=Matrix(SIZE − 4) :*
>   *# This bit of code just forms a display of the top right hand corner of the matrix*

> *#for pp from 1 to (SIZE − 4)* **do**

> *#for qq from 1 to (SIZE − 4)* **do**

> *#JACK[pp, qq] := ANDY[pp, qq + shift]*

> *#end do: end do:*

> *#JACK:=JACK;*

>

> *#PROD:=evalm(ANDY&*SYD) :*
>   *# This bit of code is of practical use in eliminating parameters by hand.*

> *#row:=4:   col:=6:*

> *#ANDY[1, op(op(VARMAT[row, col])) + 1];*

> *#for zz from 1 to SIZE do*

> *#factor(select(has, expand(PROD[row, col] − P[row, col] \cdot PROD[1, op(op(VARMAT[row, col])) + 1]), A[zz]));*

> *#end do;*

Stop Press: Yet one more attempt to calculate the algebraic matrix form has resulted in a multiplicatively closed algebraic matrix form with only ten free parameters. This elimination procedure ended in a single quadratic equation as did the code above. The algebras given by the second attempt are the same as those given by the attempt shown above. Perhaps it is the nature of the order twelve A_4 group that there is no eleven parameter algebraic matrix form. There is much not known about this area of mathematics.

Chapter 24

Concluding Remarks

Your author began this exploration of the division algebras of the groups of order five to fifteen seeking groups other than the $C_2 \times C_2$ that held geometric spaces. We have found no such other geometric spaces in the groups we have been able to examine.

By observation of the universe about us, we might have known that the only two geometric spaces that exist are the 4-dimensional space-time with distance function $d^2 = t^2 - x^2 - y^2 - z^2$, which is the space of classical physics, and the quaternion geometric space with distance function $d^2 = w^2 + x^2 + y^2 + z^2$ which is the space of the physicist's spinors (pairs of Euclidean complex numbers). None-the-less, we must always check our understanding against whatever mathematical or physical experiments we can construct.

Our exploration of the division algebras within the groups of order five to fifteen has not been as thorough or as complete as we would have wished it to be. We have been stymied by lack of computer power and by our own lack of understanding of which groups will hold which division algebras and why they hold those division algebras rather than other division algebras. We have had surprises. We would not have imagined that the order eight dicyclic group, the quaternion group, held 128 copies of only one division algebra. We might have expected that the group $C_3 \times C_3$ would hold non-commutative algebras that supported geometric spaces as the group $C_2 \times C_2$ does. We have found unexpected variety in the way that division algebras are distributed within the different groups.

Of course, we have only just begun to explore this newly discovered continent of mathematics. We have come to the foothills of our first mountain range and been unable to ascend any further than a few footsteps. We are very aware that our weakness of understanding is holding us back just as humanity has been initially held back every time we have found a new frontier and ventured into the unknown beyond that frontier. Your author does not doubt that many more able mathematicians than himself will open this new continent of mathematics by their endeavour. Your author leaves this exploration to them for he has other fish to fry.

From the work reported in this book, your author wanted to find evidence to support his contention that the space-time classical physics and quantum physics of our universe derive from no more than the finite groups and in particular from the finite group $C_2 \times C_2$. Your author is aware that the group $C_2 \times C_2$ alone, although holding all classical physics and seemingly much quantum physics, seems not to hold the strong force, and he hoped to catch a glimpse of the origin of the strong force during this exploration of groups from order five to fifteen. We have seen no glimpse of the strong force, but we have seen much evidence for the uniqueness of our 4-dimensional universe.

We have been able to show that the very great majority of finite groups do not hold geometric spaces. We have the result:

a) Only groups of order 2^n might hold geometric spaces.

b) Groups with one or more C_4 sub-groups do not hold geometric spaces

This is sufficient to eliminate all finite groups except $C_2 \times C_2$ & C_2. The extra conditions just add weight.

c) All groups with cyclic subgroups of order three to fifteen do not hold geometric spaces.

d) All groups of prime order do not hold geometric spaces.

e) All groups of odd order do not hold geometric spaces

f) All groups will odd ordered subgroups do not hold geometric spaces

g) All symmetric groups, since they have a S_3 subgroup, do not hold geometric spaces.

h) Any group that has $C_2 \times C_2 \times C_2$ or $C_3 \times C_3$ or the quaternion group as a subgroup do not hold geometric spaces.

Without doubt (Sylow's theorems) groups of order 2^n will have a C_4 sub-group and are thus prevented from holding a geometric space. We have found that no cyclic group, and hence no group which as proper cyclic subgroups, holds a geometric space.

If mathematics was an observational science, we would have verified with a very high sigma level that the only geometric spaces which exist are those that derive from the $C_2 \times C_2$ group and which we observe to form our universe. Of course, mathematicians demand a much higher level of proof than mere concordance with observation.

Your author's main interests are in physics. The next step for your author is to seek to derive quantum theory from the quaternion geometric space. He has not the time to dwell upon this new continent of mathematics but must vacate it for younger mathematicians to explore. Perhaps he will return in quest of the strong force. Perhaps he will return with a more powerful computer at some time in the future. For now, this new continent of mathematics is for the young to conquer. If you are such a young explorer, please take my blessing with you. I wish you much success and promise you much surprise and awe.

It has been a pleasure writing for you.

Dennis Morris

Port Mulgrave

Christmas Day 2015.

Other Books by the Same Author

The Naked Spinor – a Rewrite of Clifford Algebra

Spinors exist in Clifford algebras. In this book, we explore the nature of spinors. This book is an excellent introduction to Clifford algebra.

Complex Numbers The Higher Dimensional Forms – Spinor Algebra

In this book, we explore the higher dimensional forms of complex numbers. These higher dimensional forms are connected very closely to spinors.

Upon General Relativity

In this book, we see how 4-dimensional space-time, gravity, and electromagnetism emerge from the spinor algebras. This is an excellent and easy-paced introduction to general relativity.

From Where Comes the Universe

This is a guide for the lay-person to the physics of empty space.

Empty Space is Amazing Stuff – The Special Theory of Relativity

This book deduces the theory of special relativity from the finite groups. It gives a unique insight into the nature of the 2-dimensional space-time of special relativity.

The Nuts and Bolts of Quantum Mechanics

This is a gentle introduction to quantum mechanics for undergraduates.

Quaternions

This book pulls together the often separate properties of the quaternions. Non-commutative differentiation is covered as is non-commutative rotation and non-commutative inner products along with the quaternion trigonometric functions.

The Uniqueness of our Space-time

This book reports the finding that the only two geometric spaces within the finite groups are the two spaces that together form our universe. This is a startling finding. The nature of geometric space is explained alongside the nature of division algebra space, spinor space. This book is a catalogue of the higher dimensional complex numbers up to dimension fifteen.

Lie Groups and Lie Algebras

This book presents Lie theory from a diametrically different perspective to the usual presentation. This makes the subject much more intuitively obvious and easier to learn. Included is perhaps the clearest and simplest presentation of the true nature of the Lie group $SU(2)$ ever presented.

The Physics of Empty Space

This book presents a comprehensive understanding of empty space. The presence of 2-dimensional rotations in our 4-dimensional space-time is explained. Also included is a very gentle introduction to non-commutative differentiation. Classical electromagetism is deduced from the quaternions.

The Electron

This book presents the quantum field theory view of the electron and the neutrino. This view is radically different from the classical view of the electron presented in most schools and colleges. This book gives a very clear exposition of the Dirac equation including the quaternion rewrite of the Dirac equation. This is an excellent introduction to particle physics for students prior to university, during university and after university courses in physics.

The Quaternion Dirac Equation

This small book (only 40 pages) presents the quaternion form of the Dirac equation. The neutrino mass problem is solved and we gain an explanation of why neutrinos are left-chiral. Much of the material in this book is drawn from 'The Electron'; this material is presented concisely and inexpensively for students already familiar with QFT.

An Essay on the Nature of Space-time

This small and inexpensive volume presents a view of the nature of empty space without the detailed mathematics. The expanding universe and dark energy is discussed.

Elementary Calculus from an Advanced Standpoint

This book rewrite the calculus of the complex numbers in a way that covers all division algebras and makes all continuous complex functions differentiable and integrable. Non-commutative differentiation is covered. Gauge covariant differentiation is covered as is the covariant derivative of general relativity.

Even Mathematicians and Physicists make Mistakes

This book points out what seems to be several important errors of modern physics and modern mathematics. Errors like the misunderstanding of rotation, the failure to teach the higher dimensional complex numbers in most universities, and the mathematical inconsistency of the Dirac equation and some casual errors are discussed. These errors are set in their historical circumstances and there is discussion about why they happened and the consequences of their happening. There is also an interesting chapter on the nature of mathematical proof within our society, and several famous proofs are discussed (without the details).

Finite Groups – A Simple Introduction

This book introduces the reader to finite group theory. Many introductory books on finite groups bury the reader in geometrical examples or in other types of groups and lose the central nature of a finite group. This book sticks firmly with the permutation nature of finite groups and elucidates that nature by the extensive use of permutation matrices. Permutation matrices simplify the subject considerably. This book is probably unique in its use of permutation matrices and therefore unique in its simplicity.

The Left-handed Spinor

This book covers the left-handed parts of mathematics which we call the chiral algebras. These algebras have CP invariance, violation of parity, and many other aspects which makes them relevant to theoretical physics. It is quite a revelation to discover that mathematics is left-handed.

The Little Book of Permutation Matrices

This is the only book available whose subject matter is permutation matrices. Permutation matrices are very simple things, but they underlie the whole of modern theoretical physics and modern algebra. A 10 year old could understand this book.

Non-commutative Differentiation and the Commutator

(The Search for the Fermion Content of the Universe)

This book develops the theory of non-commutative differentiation from the fundamentals of algebra. We see what an algebraic operation (addition, multiplication) really is, and we discover that the commutator is a third fundamental algebraic operation within some division algebras. This leads to the first part of the derivation of the fermion content of the universe.

Index

www.ingramcontent.com/pod-product-compliance
Lightning Source LLC
Chambersburg PA
CBHW081150180526
45170CB00006B/2008